MW00965310

Treating Dense Nonaqueous-Phase Liquids (DNAPLs)

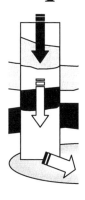

Remediation of Chlorinated and Recalcitrant Compounds

EDITORS
Godage B. Wickramanayake,
Arun R. Gavaskar, and Neeraj Gupta
Battelle

The Second International Conference on Remediation of Chlorinated and Recalcitrant Compounds

Monterey, California, May 22–25, 2000

BATTELLE PRESS
Columbus • Richland

Library of Congress Cataloging-in-Publication Data

International Conference on Remediation of Chlorinated and Recalcitrant Compounds
(2nd : 2000 : Monterey, Calif.)
    Treating dense nonaqueous-phase liquids (DNAPLs) : remediation of chlorinated and
recalcitrant compounds (C2-2) / edited by Godage B. Wickramanayake, Arun R.
Gavaskar, Neeraj Gupta.
        p.   cm.
    "The Second International Conference on Remediation of Chlorinated and
Recalcitrant Compounds, Monterey, California, May 22–25, 2000."
    Includes bibliographical references and index.
    ISBN 1-57477-096-9 (alk. paper)
        1. Dense nonaqueous phase liquids--Environmental aspects--Congresses. 2.
Organochlorine compounds--Environmental aspects--Congresses. 3. Hazardous waste
site remediation--Congresses. I. Wickramanayake, Godage B., 1953– II. Gavaskar,
Arun R., 1962– III. Gupta, Neeraj, 1962– IV. Title.

TD1066.D45 I58 2000
628.5'2--dc21

                                                            00-034239

Printed in the United States of America

Battelle Press
505 King Avenue
Columbus, Ohio 43201-2693, USA
614-424-6393 or 1-800-451-3543
Fax: 1-614-424-3819
E-mail: press@battelle.org
Website: www.battelle.org/bookstore

For information on future environmental remediation meetings, contact:
Remediation Conferences
Battelle
505 King Avenue
Columbus, Ohio 43201-2693, USA
Fax: 614-424-3667
Website: www.battelle.org/conferences

CONTENTS

DNAPL Modeling

DNAPL Remediation Technologies

DNAPLs in Fractured Media

Surfactant/Cosolvent Flushing

FOREWORD

Dense nonaqueous-phase liquids (DNAPLs) pose daunting challenges to environmental remediation professionals at contaminated sites throughout the world. Unfortunately, DNAPL sources typically are difficult to locate and characterize, persistent in their ability to contaminate soil and groundwater, and stubborn in their resistance to remediation. *Treating Dense Nonaqueous-Phase Liquids (DNAPLs): Remediation of Chlorinated and Recalcitrant Compounds* combines theoretical and practical approaches, providing information on recent advances in DNAPL site characterization, DNAPL modeling, DNAPL remediation technologies, DNAPLs in fractured media, and surfactant/cosolvent flushing.

This is one of seven volumes resulting from the Second International Conference on Remediation of Chlorinated and Recalcitrant Compounds (May 22–25, 2000, Monterey, California). Like the first meeting in the series, which was held in May 1998, the 2000 conference focused on the more problematic contaminants—chlorinated solvents, pesticides/herbicides, PCBs/dioxins, MTBE, DNAPLs, and explosives residues—in all environmental media and on physical, chemical, biological, thermal, and combined technologies for dealing with these compounds. The conference was attended by approximately 1,450 environmental professionals involved in the application of environmental assessment and remediation technologies at private- and public-sector sites around the world.

A short paper was invited for each presentation accepted for the program. Each paper submitted was reviewed by a volume editor for general technical content. Because of the need to complete publication shortly after the Conference, no in-depth peer review, copy-editing, or detailed typesetting was performed for the majority of the papers in these volumes. Papers for 60% of the presentations given at the conference appear in the proceedings. Each section in this and the other six volumes corresponds to a technical session at the Conference. Most papers are printed as submitted by the authors, with resulting minor variations in word usage, spelling, abbreviations, the manner in which numbers and measurements are presented, and formatting.

We would like to thank the people responsible for the planning and conduct of the Conference and the production of the proceedings. Valuable input to our task of defining the scope of the technical program and delineating sessions was provided by a steering committee made up of several Battelle scientists and engineers – Bruce Alleman, Abraham Chen, James Gibbs, Neeraj Gupta, Mark Kelley, and Victor Magar. The committee members, along with technical reviewers from Battelle and many other organizations, reviewed more than 600 abstracts submitted for the Conference and determined the content of the individual sessions. Karl Nehring provided valuable advice on the development of the program schedule and the organization of the proceedings volumes. Carol Young, with assistance from Gina Melaragno, maintained program data, corresponded with speakers and authors, and compiled the final program and abstract books. Carol and Lori Helsel were responsible for the proceedings production effort, receiving assistance on specific aspects from Loretta Bahn, Tom Wilk, and Mark Hendershot. Lori, in particular, spent many hours examining papers for format and contacting authors as necessary to obtain revisions. Joe Sheldrick, the manager of Battelle Press, provided valuable production-planning advice; he and Gar Dingess designed the volume covers.

Battelle organizes and sponsors the Conference on Remediation of Chlorinated and Recalcitrant Compounds. Several organizations made financial contributions toward the 2000 Conference. The co-sponsors were EnviroMetal Technologies Inc. (ETI); Geomatrix Consultants, Inc.; the Naval Facilities Engineering Command (NAVFAC); Parsons Engineering Science, Inc.; and Regenesis.

As stated above, each article submitted for the proceedings was reviewed by a volume editor for basic technical content. As necessary, authors were asked to provide clarification and additional information. However, it would have been impossible to subject more than 300 papers to a rigorous peer review to verify the accuracy of all data and conclusions. Therefore, neither Battelle nor the Conference co-sponsors or supporting organizations can endorse the content of the materials published in these volumes, and their support for the Conference should not be construed as such an endorsement.

Godage B. Wickramanayake and Arun R. Gavaskar
Conference Chairs

USE OF THE RAPID OPTICAL SCREENING TOOL TO EVALUATE DNAPL EXTENT

Alison Jones and *Frank Szerdy*
(Geomatrix Consultants, Inc., Oakland, CA)
Mitchell Brourman and Michael Slenska
(Beazer East, Inc., Pittsburgh, Pennsylvania)
Andrew Taer (Fugro Geosciences, Inc., Houston, Texas)

ABSTRACT: The authors are designing subsurface barriers at two wood-treatment facilities to physically contain dense non-aqueous phase liquids (DNAPLs) in the subsurface. A combination of conventional soil borings and a Rapid Optical Screening Tool (ROST^{TM}) deployed on a Cone Penetration Testing (CPT) probe was used to evaluate the DNAPL extent. A correlation between the ROST^{TM} response and data from soil samples was developed. Using these data, the extent of DNAPL in the subsurface was estimated and zones of free phase product were differentiated from zones where DNAPL was present, but not so pervasive. It is suggested that the derived correlation may be useful at other sites as an interpretative tool.

INTRODUCTION

The authors are designing subsurface barriers at an operational wood-treatment facility in South Carolina (Site 1) and at a former wood treatment facility in Maryland (Site 2). The purpose of these barriers is to physically contain and isolate dense non-aqueous phase liquids (DNAPLs) in the subsurface that represent long-term sources of dissolved phase constituents to groundwater; creosote is the DNAPL of primary concern.

Field investigations were completed to evaluate the extent of DNAPL along the proposed barrier alignments, to develop stratigraphic information along the proposed alignments, and to obtain sufficient geotechnical data to complete the barrier designs. These data were used to define the specific alignment of each barrier. A combination of conventional soil borings and Fugro Geosciences' Rapid Optical Screening Tool (ROST^{TM}) deployed on a Cone Penetration Testing (CPT) probe was utilized to obtain the required information.

FIELD INVESTIGATION METHODS

Soil Borings. Soil borings were drilled using hollow-stem auger and mud-rotary methods. Standard penetration testing (SPT) samples were collected according to ASTM standard D1586-84 approximately every 2 feet (0.6 meters). Soil samples were observed and logged in the field and screened for the presence of DNAPL by visual and olfactory identification.

Cone Penetration Testing (CPT). The CPT is a proven method for rapidly evaluating the physical characteristics of unconsolidated soils. It is based on the resistance to penetration of an electronically-instrumented cone which is continuously advanced into the subsurface. The standard geotechnical sensors within the cone measure tip resistance and sleeve friction in tons per square foot (TSF). The combined data from the tip resistance and sleeve friction form the basis for soil classification. Soil stratigraphy was identified using empirical correlations developed by Robertson and Campanella (1986).

The CPT is a method of generating lithologic data without generating potentially contaminated or hazardous cuttings. Information is generated rapidly and the CPT can be a cost-effective method of investigation when compared to soil borings. However, some soil borings are needed to calibrate the CPT data.

In accordance with ASTM standard D5778-95, the cone was advanced at a rate of 2 centimeters per second with the driving force provided by hydraulic rams. The CPT was operated from a Fugro Geoscience, customized, 25-ton CPT truck. The CPT cone used had an apex angle of 60 degrees, a base area of 15 square centimeters (cms^2), and a friction sleeve area of 200 cms^2.

ROST^{TM} Testing. The Fugro Geosciences ROST^{TM} laser-induced fluorescence system was used to screen soils for DNAPL. The system consists of a tunable laser mounted in the CPT truck that is connected to a down-hole sensor. The down-hole sensor consists of a small diameter, sapphire window mounted flush with the side of the cone penetrometer probe. The laser and associated equipment transmit 50 pulses of light per second to the sensor through a fiber optic cable.

The laser light passes through the sapphire window and is absorbed by aromatic hydrocarbon molecules in contact with the window as the probe is advanced. This addition of energy (photons) to the aromatic hydrocarbons causes them to fluoresce. A portion of the fluorescence emitted by any encountered aromatic constituents is returned through the sapphire window and conveyed by a second fiber optic cable to a detection system within the CPT rig. The emission data resulting from the pulsed laser light is averaged into one reading per second interval (approximately one reading per 2 cms vertical interval) and is recorded continuously. The ROST^{TM} may be operated in single or multi-wavelength mode; at both Site 1 and Site 2 the ROST^{TM} was operated in multi-wavelength mode (MWL).

In MWL mode, the emitted fluorescence is measured simultaneously at four monitoring wavelengths (340, 390, 440 and 490 nm). The four monitoring wavelengths cover the range of light produced by light fuels though heavy contaminants such as coal tar and creosote and enhance detection of widely ranging product types. The emission data are reported continuously as a total of the fluorescence intensity recorded at each of the four wavelengths. The total fluorescence intensity data are presented in real-time on a computer monitor as a graph of fluorescence intensity versus depth. Fluorescence intensity is recorded relative to a standard reference solution, considered to have a fluorescence intensity of 100 percent.

The relative percentage of fluorescence measured at each of the four monitoring wavelengths is plotted continuously on the ROST^{TM} logs as four continuous "color bands". The width of each color band represents the relative percentage of fluorescence emitted by the contaminant at each of the monitoring wavelengths. For general interpretation purposes, lighter aromatic hydrocarbon molecules will emit fluorescence at shorter wavelengths and heavier, longer chained hydrocarbons will emit fluorescence at longer wavelengths.

ROST^{TM} Calibration. At both Site 1 and Site 2, creosote is the DNAPL of primary concern. Before starting each field program, the ROST^{TM} probe was inserted in a bucket of pure creosote to identify the representative ROST^{TM} response for the product at these sites. These data are presented on Figure 1. It may be noted that the maximum fluorescence intensity of this product (relative to the reference solution) is somewhere between 5 and 6 percent and that peak fluorescence is in the 440 to 490 nm range. The product waveform is typical of creosote.

Site Specific Field Programs. To obtain the required information in the most cost-effective manner, CPTs and borings were located at spacings of between 100 and 250 feet. At Site 1, where the length of the subsurface barrier will be approximately 2,500 feet long (760 m), 8 borings and 16 CPTs were completed. Of the 16 CPTs, 7 were completed adjacent to soil borings for calibration purposes making the average spacing between information points approximately 150 feet (46m). In the field, penetration to the required depth (approximately 70 feet below ground surface, bgs) with the CPT proved to be problematic because of very dense sands identified at certain locations. Thirteen CPTs reached refusal at depths of between 40 and 50 feet bgs, approximately 5 feet into dense sand with SPT blowcounts in excess of 100 blows/foot. As a result, more soil borings and fewer CPTS were completed than originally planned.

At Site 2, where the length of the subsurface barrier will be approximately 5,750 feet long (1,750 m), 9 borings and 25 CPTs were completed. Of the 25 CPTs, 3 were completed adjacent to soil borings for calibration purposes making the average spacing between information points approximately 185 feet (56m). Even though very dense sands (blowcounts in excess of 50 blows/foot) were encountered at depths of approximately 30 feet bgs, most of the CPTs penetrated to the required investigation depth of approximately 55 feet bgs.

RESULTS AND DISCUSSION

Site 1 Stratigraphy. Based on previous investigations, two water-bearing units have been identified beneath Site 1. The upper water-bearing unit is an unconfined aquifer identified in the project documents as the A and B-zones. The A-zone lithology consists of silty clay to clayey sand. The B-zone consists primarily of poorly graded sands. The upper A and B-zone aquifer is separated from the second water-bearing unit, the C-zone, by a clay rich aquitard (the B/C aquitard). The B/C aquitard lithology varies from interbedded poorly graded sand and lean clay to lean clay. Site data indicate that the DNAPL is limited to the A and B-zones. It is proposed to key-in the subsurface containment barrier to the clay-rich B/C aquitard.

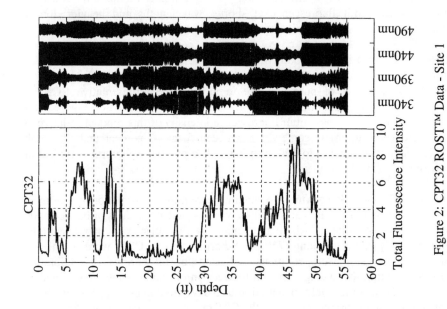

Figure 2: CPT32 ROST™ Data - Site 1

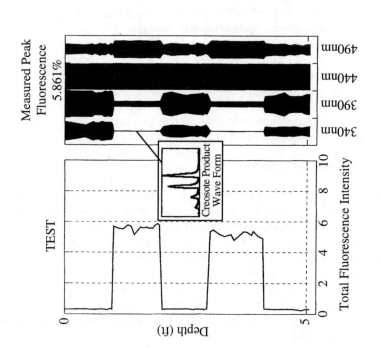

Figure 1: Test ROST™ Data - Pure Creosote

Site 2 Stratigraphy. Based on previous investigations, the Site 2 geology has been characterized as consisting of three zones: A, B-C and D. The A-zone is generally 20 to 30 feet thick and soil types are typically variable. Generally, the A-zone includes approximately 5 feet of poorly graded sand fill over varying amounts of organic clay and peat over poorly graded sands and silt. Below approximately 20 to 30 feet bgs, the stratigraphy of the B/C zone consists of medium dense to very dense sands. The B/C and D zones are separated by a regional confining layer (the C/D aquitard). Along the proposed subsurface barrier alignment, the depth to the top of the C/D aquitard varies between 40 and 57 feet bgs; the aquitard is characterized as lean clay to silt. It is proposed to key-in the subsurface containment barrier to the clay-rich C/D aquitard.

Site 1 ROST™ Data. At both sites, the purpose of the proposed subsurface barriers is to physically contain and isolate dense non-aqueous phase liquids (DNAPLs) in the subsurface. At Site 1, the obstacles to the construction of the subsurface barrier include railway tracks to the south and east. The intent is to construct the wall beyond the limits of the free-phase product within these physical constraints. With this in mind, the borings and CPT/ROST™ probes were located where existing data indicated the limits of the free-phase DNAPL might reasonably be expected. It was not anticipated therefore, that significant amounts of free-phase DNAPL would be identified during the investigation. However, a single CPT/ROST™ probe was completed at a location within the proposed subsurface barrier known to contain significant amounts of free-phase DNAPL. These data are presented on Figure 2. ROST™ responses more typical of the proposed barrier alignment, show very little response above baseline fluorescence (attributed to the native soil matrix) especially at depth. Space limitations preclude the inclusion of additional examples here.

Correlation Between ROST™ and Soil Boring Data. Correlations between ROST™ fluorescence intensity and petroleum hydrocarbon concentrations in soil have been established at other sites. At another creosote site, Ruggery et al., (1996) identified a high degree of correlation between total hydrocarbon concentrations in soil and ROST ™ fluorescence intensity. Data from other sites also indicate that ROST™ fluorescence intensity is generally proportional to petroleum hydrocarbon concentrations in soil (Taer et al. 1995) although Taer et. al do state that, due to the heterogeneity of the soil matrix, ROST™ fluorescence intensity should not be relied upon to accurately quantify total hydrocarbon concentrations.

At Sites 1 and 2, there was less concern with identifying DNAPL concentrations in the soil than with identifying the presence of free-phase product. It was decided that the cost of obtaining analytical data was unwarranted and that sufficient design information could be obtained from soil borings and CPT/ROST™ probes. However, in order to fully define the extent of the DNAPL in the subsurface, a method of correlating the ROST™ response with observable data from soil samples was needed. A way to represent the two sets of data in a consistent manner was also needed.

By comparing soil and ROST™ data from soil borings and CPT probes located adjacent to each other, a correlation between the two sets of data was established and a consistent method of interpreting, or defining, these data sets was established. This correlation is presented on Table 1; data from Site 1 were used to establish the correlation.

TABLE 1. Creosote: An Approximate Correlation Between the ROST™ Response and Observable Data from Soil Samples

ROST™ Response	Soil Data	Interpretation
Above background to 1% intensity	Creosote odor but no other observable indication	Weak response
1 to 3% intensity	Product staining to creosote in stringers/fractures	Medium response
3 to 6% intensity	Free product	Strong response

Identification of DNAPL at Site 1. Stratigraphical cross-sections were developed that show site geology and indications of DNAPL. At small scale, and in black-and-white, these cross-sections are not particularly illuminating and are not therefore presented here. Each ROST™ fluorescence intensity versus depth profile was represented graphically as zones of no, weak, medium and strong response. Each borehole log was represented similarly. Where a creosote odor had been observed, the response was represented as weak. Where no odor, or any other indication of creosote was observed, the graphical representation was of no response.

Based on these cross-sections, and additional data from existing monitoring wells, the extent of DNAPL in the subsurface was estimated and zones of free phase product were differentiated from zones where DNAPL was present, but not so pervasive.

Zones of DNAPL identified from the soil borings, monitoring wells, and ROST™ data were compared with historical aerial photographs of the site. Zones of free product identified from the soil borings, monitoring wells and ROST™ data coincided closely with former waste lagoons and other features where DNAPL is known to have been present. These data were presented graphically for different depths. An example for the 20 to 40 feet bgs depth range is presented on Figure 3. These data, in combination with data from other depths, were used to finalize the subsurface barrier alignment at Site 1. Data for the 0 to 20 feet bgs depth range showed a strong response encompassed by the northern portion of the subsurface barrier.

Identification of DNAPL at Site 2. Site 2 lies on a major river. It has been agreed with the appropriate regulatory agencies that the subsurface barrier will be

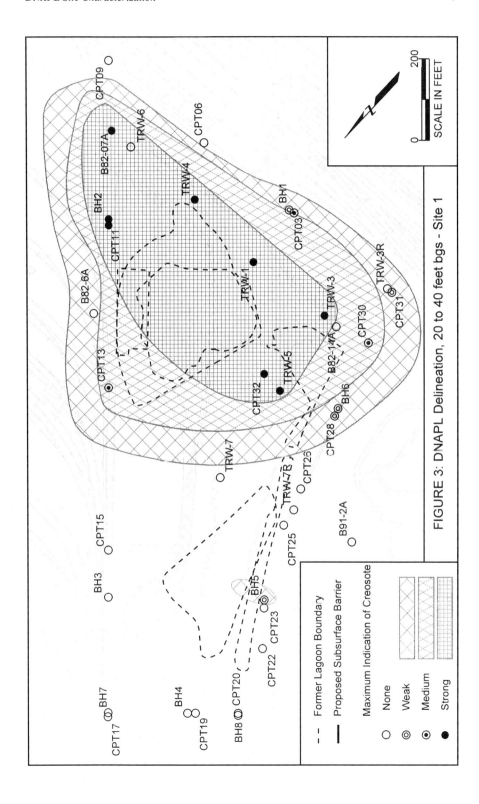

FIGURE 3: DNAPL Delineation, 20 to 40 feet bgs - Site 1

constructed to contain as much of the free-phase DNAPL as is possible without encroaching on the river. As a result, in some localized spots, there will be some free-phase product that is not fully contained by the subsurface barrier. ROST™ data collected along the proposed subsurface barrier alignment at Site 2, showed a greater range in intensities than was observed at Site 1. These data confirmed that the subsurface barrier will go through some localized areas containing free-phase product.

The ROST™ and soil data obtained at Site 2 were interpreted using the correlation between the ROST™ response and observable data from soil samples presented on Table 1. These data were presented graphically on stratigraphic cross-sections. A direct comparison between the two sets of data indicated that the correlation between ROST™ and observable data presented on Table 1 was also applicable at Site 2. This indicates that the correlation derived from Site 1 data may be useful as an interpretative tool at other sites.

CONCLUSIONS

1. In combination with a limited number of soil borings, the CPT can be a cost-effective method of generating lithologic data without generating potentially contaminated or hazardous cuttings. However, the CPT is not an appropriate tool where very dense sands are present at depth.
2. The ROST™ can be a cost-effective method of quickly identifying and delineating DNAPLs in the subsurface.
3. The correlation presented in this paper between the ROST™ response and observable data from soil samples developed using data from Site 1 proved to be applicable at Site 2. The same correlation may prove useful at other creosote sites as an interpretative tool and as a method of presenting data from soil borings and the ROST™ in a consistent manner.

REFERENCES

Robertson, P.K., and R.G. Campanella. 1986. "Guidelines for Use, Interpretation and Application of the CPT and CPTU." *Hogentogler and Company.*

Ruggery, D.A., N.J. Misquitta, and F.R. Coll. 1996. "Delineation of Creosote-Based DNAPLS Using CPT-Deployed Laser Induce Fluorescence," *The Sixth West Coast Conference on Contaminated Soils and Groundwater, Newport Beach, California.*

Taer, A. D., R.F. Farrell, and B. Ford. 1995. "Use of the Cone Penetrometer and Laser-Induced Fluorescence in an Oil Field Site Characterization,"

PERFORMANCE ASSESSMENT OF IN-SITU REMEDIAL OPERATIONS INVOLVING NAPLS

Richard E. Jackson, John E. Ewing, Minquan Jin, and Hans W. Meinardus
(Duke Engineering & Services, Austin, Texas)

ABSTRACT: There is no generally accepted protocol for the performance assessment (PA) of remedial operations that remove chlorinated degreasing solvents, fuel hydrocarbons and other NAPLs from the subsurface. Many PA studies simply rely upon the change in aqueous concentration of the contaminant of concern in ground water before, during and after remediation. This assumes an adjacent NAPL zone and that the changes in ground-water concentrations reflect proportional changes in NAPL mass and/or volume. Frequently PA studies have used soil cores to demonstrate changes in the mass of the contaminant of concern before, during and after remediation. This approach assumes that enough soil cores are collected without loss of NAPL, preserved and analyzed to sample the representative elementary volume of the NAPL zone. A third approach is to conduct partitioning interwell tracer tests (PITTs) to measure not only how much NAPL was present before remediation but also how much remained afterwards. This method assumes that the PITT sweeps even low permeability units containing NAPL when high-permeability 'thief' zones may offer preferential transport pathways for the tracers. The assumptions underlying each approach are examined.

INTRODUCTION

NAPL removal is occurring at numerous sites across the USA in the absence of clear guidelines for assessing the performance of the removal operations, a condition noted by the Committee on Innovative Remediation Technologies of the National Research Council (1997). Because of the unique nature of each NAPL-contaminated site and the nature of the various remedial processes employed, the design of PA schemes varies from site to site. Generally, the US EPA does not prescribe the data requirements for PA; it merely identifies general approaches to data collection and its quality control. This situation has led to the common occurrence in which the performance of the remedial operations to remove NAPL has been defined by the vendor rather than any disinterested party. The conflict of interest in both undertaking and assessing remedial performance is clear and general principles of practice are obviously required.

This reluctance to prescribe precise guidelines has resulted in a wide variety of practices, which are summarized in Table 1. This Table is composed of information taken from the volumes of the First International Conference on Remediation of Chlorinated and Recalcitrant Compounds held in Monterey, California in May 1998.

TABLE 1. Methods of performance assessment reported in the Proceedings of the First International Conference on Remediation of Chlorinated and Recalcitrant Compounds, 1998, Battelle Press, Columbus, OH.

| In-Situ Remedial Technology | Contaminants | PA Methods | | | First Author, volume & page |
		Soils	GW	PITT	
Air Sparging	DCE, PCE, TCE		✓		Hughes, 1(5):279
Air Sparging	DCE, PCE, TCE		✓		Brown, 1(5):285
Bioremediation	PCE, TCE		✓		Legrand, 1(4):193
Bioremediation	DCE, PCE, TCE		✓		Dacyk, 1(4):215
Bioremediation	TCA, TCE, CHCl₃		✓		Cox, 1(4):227
Bioremediation	PCE		✓		Becvar, 1(4):121
Cosolvent Flooding	Multi-component LNAPL	✓	✓	✓	Falta, 1(2):205
Hydrous Pyrolysis & Oxidation + Steam	Creosote DNAPL		✓		Leif, 1(5):133
Oxidation	TCE	✓	✓		McKay, 1(5):377
Oxidation	PCE-TCE DNAPL	✓			Jerome, 1(5):353
Oxidation	PCP, PAHs	Soil gas	✓		Marvin, 1(5):383
Oxidation	TCE	✓			Levin, 1(5):437
Surfactant flooding	TCE DNAPL	✓	✓	✓	Meinardus, 1(2):143
Steam stripping	TCE, DCE	✓	✓		Rice, 1(5):127
Thermal Desorption	PCBs	✓			Vinegar, 1(5):25
Vitrification	PCBs	✓			Campbell, 1(6):231

Terminology Used: Soils = soil sampling; GW= ground-water sampling.

Table 1 suggests that ground water is the preferred phase to measure for air sparging and bioremediation, although NAPL may or may not have been present in the zones under remediation. However, an inspection of the various articles reveals little or no evidence of a statistical design to the sampling of the monitoring wells. When the remediation zone is known or suspected to contain NAPL, soil sampling (again apparently without some statistical design protocol) is undertaken even though the heterogeneity of the porous medium may make it very difficult to obtain a representative sample. In those cases where NAPL removal technologies are employed, e.g., hydrous pyrolysis, chemical oxidation, steam stripping and surfactant flooding, soil sampling is used for PA but not where NAPL is assumed to be absent. The partitioning interwell tracer test (PITT) is also used to measure the volume of NAPL and its spatial distribution before and after NAPL removal, particularly in the case of chemical floods involving surfactants and cosolvents. The PITT involves the injection and subsequent extraction of a tracer solution into the NAPL removal zone and the measurement of the breakthrough curves at the extraction wells, which are then interpreted to determine the volume of NAPL in the interwell zone (Jin et al., 1995).

The purpose of this presentation is to review the assumptions implicit in the use of soil cores, ground-water samples and PITTs to assess the performance of NAPL removal operations. The context of this paper is a typical geosystem composed of ground water, NAPL, alluvial aquifer material and, perhaps, capillary barriers preventing NAPL migration.

PA BY SOIL SAMPLING

The collection of soil cores, taken before and after a remedial operation, raises three issues of great consequence for the role of soil sampling in NAPL remediation:

1. Do the samples represent average values for that part of the subsurface that they are chosen to represent? That is, do the samples meet the requirement that they constitute NAPL concentration measurements for the representative elementary volume of that part of the geosystem?
2. Has the method of the recovery and handling of the soil core resulted in the loss of significant amounts of NAPL?
3. Have soil samples been recovered from all parts of the NAPL zone under remediation, in particular parts of the geosystem that have relatively high and relatively low permeabilities?

The first issue has been addressed by Mayer and Miller (1992), who have shown that the scale of measurement for residual NAPL (i.e., NAPL held by capillary forces) in aquifers is probably much larger than a typical soil or aquifer sample. They reported that as the porous medium becomes more non-uniform, the necessary volume of aquifer to be sampled to yield a representative value of average NAPL saturation (percentage of pore volume occupied by residual NAPL) increased rapidly. Therefore, when NAPL is the subject of in-situ remediation, very large numbers of soil cores, each of a large volume, are required. Stan Feenstra stated at the recent Theis Conference in Florida that he felt that the vertical sampling frequency for soil cores as indicated by the experiments conducted at Canadian Forces Base Borden was every 5 cm (2 in). Our own work at Hill AFB suggests that the required frequency is between 5 and 15 cm. Assuming that such a requirement is met, one can either compare the average pre- and post-remediation soil concentrations of several target compounds or interpolate between the various representative core samples and determine the volume of NAPL represented by these cores. Because soil cores are a form of destructive sampling, it is impossible to obtain pre- and post-remediation samples from the same location and this method assumes that the average concentrations are representative of the entire geosystem. Given the sparse (areally) nature of soil core data, the process of interpolation involves degrees of freedom in the statistical method and parameters employed which result in a wide array of possible answers for the NAPL volume and distribution. Thus, in the words of the SERDP/EPA group that designed the Hill AFB OU1 demonstrations, soil samples "do not give good estimates of total mass in the swept volume being remediated" (Bedient et al., 1999).

The second issue is equally challenging because of the volatile nature of many NAPLs, particularly chlorinated degreasing solvents, which may be recovered by soil sampling. Those who have observed soil coring operations in sandy aquifer materials know full well the difficulty of recovering the aquifer material without loss of the interstitial fluids during core recovery. This loss is an inevitable result of air entry into the core sample that causes NAPL to drain out of the sample during core recovery. Only the successful use of cohesion-less soil samplers or other coring tools that inhibit air entry into the core can minimize this problem. The NAPL drainage is due to the much lower average NAPL saturations that occur under three-phase conditions (i.e., air-water-NAPL) than under two-phase (water-NAPL) conditions (see Mercer and Cohen's (1990) Table 3). In addition, there is the further problem of the volatilization of the NAPL during sample handling at the drilling site and en route to the laboratory for analysis. The work of Hewitt et al. (1995) has shown how important proper field preservation is for all volatile components of a soil sample.

The third assumption concerns the need to collect representative soil samples from all parts of the geosystem under remediation. It is oftentimes difficult to recover soil cores from cohesion-less sandy sediments that might comprise the larger part of a contaminated aquifer. These permeable sediments would be relatively accessible to various injected surfactants or steam or oxidants, whereas finer grained aquifer materials are more readily recovered by soil sampling methods. The assessment of the risk from NAPL remaining following remediation requires knowledge of the volume and/or mass left unremoved in all parts of the former NAPL zone.

PA BY GROUND-WATER SAMPLES

As Table 1 shows, this approach to PA is the most popular, no doubt because it is also the least expensive. It makes two principal assumptions:

1. There is a unique relationship between the mass and/or volume of NAPL in an aquifer and the aqueous concentration of a dissolved component of that NAPL at an adjacent monitoring well.
2. The monitoring well is screened such that it samples stream tubes that have passed through the NAPL zone and thus records changes in NAPL mass and/or volume solely due to the remedial operations.

The use of ground-water contaminant concentrations to draw conclusions about either the volume, mass, or saturation of the NAPL zone involves an ill-posed inverse problem for which no unique solution exists even for the simplest of cases. The problem is essentially irresolvable because of the dependence of the dissolved-phase concentration on the average NAPL saturation (L^3/L^3), the specific surface area (L^2/M) of the NAPL, and the aspect ratio between the monitoring well and the NAPL zone, all of which can vary significantly during remediation. Complicating the issue even further is the change in composition of a multi-component NAPL as more soluble components are preferentially dissolved. These two assumptions are addressed in Jackson and Mariner (1995).

PA BY PITTS

The partitioning interwell tracer test has been used to assess the performance of remedial operations involving NAPL removal. Typically, it is used before remediation to assist in the design of a surfactant or cosolvent flood and immediately after the flood to determine the volume and spatial distribution of the NAPL remaining. The use of the PITT implies that:

1. the partition coefficients for the various tracers with respect to the NAPL have been measured accurately;
2. the retardation of the partitioning tracers is due solely to partitioning into and from NAPL rather than by reversible sorption to sedimentary organic matter;
3. the tracer solution penetrates and reversibly partitions with all parts of the NAPL zone, even parts of the zone containing free-phase NAPL or of low intrinsic permeability or relative permeability, i.e., due to high NAPL saturations;
4. the tracer solution is composed of at least three partitioning tracers with partition coefficients that vary over an order of magnitude to compensate for the unknown (to the design engineer) average saturations and permeabilities of the geosystem; and
5. the PITT is conducted and monitored over sufficient time that the tracer signals from all parts of NAPL zone are measured in the tracer breakthrough curves at the monitoring or extraction wells.

The design, execution and analysis of PITTs have been the subjects of a series of papers that have addressed these assumptions. The accuracy of the PITT is most sensitive to errors in the partition coefficient that may be due to poor analytical methodology or an absence of free-phase NAPL with which to undertake the measurements. For the case in which the NAPL exists only as trapped, immobile liquid, i.e., no NAPL can be recovered from wells, the chemical analysis of preserved soil samples is used to reconstruct the NAPL composition using the software code NAPLANAL (Mariner et al., 1997; may be downloaded from www.dnapl.com). Partition coefficients can then be estimated through their equivalent alkane carbon number (EACN; Dwarakanath and Pope, 1998) or measured using an analog NAPL comprising the proportions indicated by NAPLANAL. Therefore, the first assumption can be met by an indirect approach.

Typical alluvial aquifer materials do contain small amounts of sedimentary organic carbon, typically of the order of 0.1% to 0.01%. Therefore, laboratory column tests of uncontaminated alluvium are performed before a PITT to determine if this natural carbon can act as an interference, in which case, a correction factor is measured (see Jin et al., 1997, p. 970). It has been our experience that only fine-grained sediments contain sufficient natural organic carbon to interfere in this manner. Consequently, this second assumption can be addressed and satisfied through the recovery of uncontaminated soil samples similar to those trapping NAPL and their testing by a laboratory column PITT.

The third assumption that forms the foundation of the use of PITTs, that the tracer solution penetrates and reversibly partitions with all parts of the NAPL zone, has been addressed explicitly by Jin et al., 1997. Tracer response curves from an alluvial aquifer at Hill AFB have shown that it is possible to measure the volume of DNAPL in a low-permeability sand overlying a clay aquiclude and underlying a sandy zone with a permeability at least one order of magnitude higher (Meinardus et al., in review). The tracer response analysis yielded accurate estimates of the average DNAPL saturation in this low permeability unit and gave similar values to those from soil cores collected from the same sand. Similarly, it is possible to measure the volume of NAPL in zones with high NAPL saturations, including the presence of free-phase NAPL, although the errors involved increase substantially. These same errors can be much reduced by water flooding the NAPL zone to residual-phase saturations.

At the last Battelle conference on the Remediation of Chlorinated and Recalcitrant Compounds, Payne et al. (1998) unwittingly showed the hazards of using just one partitioning tracer. Their analysis of a three-zone geosystem in which only one zone contained NAPL, that of the lowest permeability, indicated that a partitioning tracer with a partition coefficient of 18 would not record the presence of the NAPL in this zone. It can be readily shown by numerical simulation that a design with a much less hydrophobic tracer (i.e., K<<18) would detect the presence of NAPL in this low permeability zone. Thus the fourth assumption of injecting a suite of tracers of differing hydrophobicities is necessary for the successful employment of PITTs.

The final assumption in the use of PITTs for PA, that the tracer signals are monitored over sufficient time that all parts of NAPL zone are measured in the tracer breakthrough curves, is critical to the accurate measurement of total NAPL volume. In geosystems with low NAPL saturations (e.g., ≤ 3%), most of the information in the tracer response curve is in the tail, i.e., at a late time in the PITT. Nelson et al. (1999) conducted a laboratory experiment in a sand tank (NAPL saturation = 1.4%) but failed to measure the tail of their tracer response curves and arrived at inaccurate results. (See rebuttal submitted to *ES&T* by Pope et al. and posted on the world wide web at www.dnapl.com/new/html). Examples of the careful measurement of tracer tails for both vadose and ground-water zone PITTs are shown in Dwarakanath et al. (1999).

CONCLUSIONS

The use of ground-water data for PA is limited by the absence of any unique relationship between the NAPL zone and the contaminant concentrations in monitoring wells. Indeed, an infinite number of different combinations of NAPL volume, saturation, composition and spatial distribution can result in the same concentration of dissolved contaminant in a monitoring well. This limitation alone precludes the use of ground-water sampling from even semi-quantitative assessment of remedial performance.

Soil cores are hindered by the fact that they are a destructive measurement technique disallowing any direct comparison to be made for assessing performance. Of the two indirect comparison methods available (point and volume-averaged measures), both are further hindered by the requirement that the sample be indicative of some REV within the geosystem and by the loss of fluids during sampling. Volume-averaged measures are subject to significant uncertainty resulting from the interpolation between sample points and require knowledge of the media porosity distribution. These limitations restrict soil core data to, at best, a semi-quantitative method of PA which is, in practicality, useful primarily in testing for "false negative" measurements by PITTs.

When improperly designed (e.g. Nelson et al., 1999), PITTs are also subject to significant error. Possible design pitfalls include the failure to use three or more tracers with sufficiently differing partition coefficients, the failure to accurately determine the partition coefficients and the failure to observe the tails of the tracer response curves. When properly designed however, PITTs provide a means of accurately and quantitatively measuring NAPL volume and assessing the performance of a remediation method. Furthermore it is possible to quantify the degree of uncertainty in the PITT results.

REFERENCES

Bedient, P. B., A. W. Holder, C. G. Enfield, and A. L. Wood. 1999. "Enhanced Remediation Demonstrations at Hill Air Force Base: Introduction." In *Innovative Subsurface Remediation*, M. L. Brusseau et al. (Eds.), ACS Symposium Series 725, pp. 36-48.

Dwarakanath, V. and G. A. Pope. 1998. "A New Approach for Estimating Alcohol Partition Coefficients between Nonaqueous Phase Liquids and Water." *Environmental Science and Technology*. 32(11): 1662-1666.

Dwarakanath, V., N. Deeds, and G. A. Pope. 1999. "Analysis of Partitioning Interwell Tracer Tests." *Environmental Science and Technology*. 3(21): 3829-3836.

Hewitt, A. D., T. F. Jenkins, and C. L. Grant. 1995. "Collection, Handling, and Storage: Keys to Improved Data Quality for Volatile Organic Compounds in Soil." *American Environmental Laboratory*. 7: 25-28.

Jackson, R. E. and P. E. Mariner. 1995. "Estimating DNAPL Composition and VOC Dilution from Extraction Well Data." *Ground Water*. 33(3): 407-414.

Jin, M., M. Delshad, V. Dwarakanath, D. C. McKinney, G. A. Pope, K. Sepehrnoori, C. E. Tilford, and R. E. Jackson. 1995. "Partitioning Tracer Test for Detection, Estimation, and Remediation Performance Assessment of Subsurface Nonaqueous Phase Liquids." *Water Resources Research*. 31(5): 1201-1211.

Jin, M., G. W. Butler, R. E. Jackson, P. E. Mariner, J. F. Pickens, G. A Pope, C. L. Brown, and D. C. Mckinney. 1997. "Sensitivity Models and Design Protocol for Partitioning Tracer Tests in Alluvial Aquifers." *Ground Water*. 36(6): 964-972.

Mariner, P. E., M. Jin, and R. E. Jackson. 1997. "An Algorithm for the Estimation of NAPL Saturation and Composition from Typical Soil Chemical Analyses." *Ground Water Monitoring & Remediation*. 17(2): 122-129.

Mayer, A. S. and C. T. Miller. 1992. "The Influence of Porous Medium Characteristics and Measurement Scale on Pore-Scale Distributions of Residual Nonaqueous-Phase Liquids." *J. Contaminant Hydrology*. 11: 189-213.

Meinardus, H. W., V. Dwarakanath, J. E. Ewing, G. J. Hirasaki, R. E. Jackson, M. Jin, J. S. Ginn, J. T. Londergan, C. A. Miller, and G. A. Pope. 1999. "Performance Assessment of DNAPL Remediation in Heterogeneous Alluvium Using Partitioning Interwell Tracer Tests." Submitted to *Environmental Science and Technology*.

Mercer, J. W., and R. M. Cohen. 1990. "A Review of Immiscible Fluids in the Subsurface: Properties, Models, Characterization and Remediation." *J. Contaminant Hydrology*. 6: 107-163.

National Research Council. 1997. *Innovations in Ground Water and Soil Cleanup: From Concept to Commercialization.* Committee on Innovative Remediation Technologies, NRC, National Academy Press, Washington D.C., pp. 240-245.

Nelson, N. T., M. Oostrom, T. W. Wietsma, and M. L. Brusseau. 1999. "Partitioning Tracer Method for the In Situ Measurement of DNAPL Saturation: Influence of Heterogeneity and Sampling Method." *Environmental Science and Technology*. 33(22): 4046-4053.

Payne, T., J. Brannan, R. Falta, and J. Rossabi. 1998. "Detection Limit Effects on Interpretation of NAPL Partitioning Tracer Tests." In G.D. Wickramanayake and R.E. Hinchee (Eds.), *Non-Aqueous Phase Liquids: Remediation of Chlorinated and Recalcitrant Compounds*, pp. 125-130. Battelle Press, Columbus, OH.

FULL-SCALE CHARACTERIZATION OF A DNAPL SOURCE ZONE WITH PITTS

Hans W. Meinardus, Varadarajan Dwarakanath, Richard E. Jackson, and
Minquan Jin (Duke Engineering and Services, Austin TX)
Jon S Ginn (Environmental Management Directorate, Hill AFB, UT)
G.Chris Stotler (URS Greiner Woodward Clyde, Denver CO)

ABSTRACT: This paper presents the implementation and results of four PITTs used to delineate a DNAPL source zone at Operable Unit 2, Hill Air Force Base. These PITTs, with a total swept pore volume of 285,000 gallons (>1 million liters), represent the first full-scale field application of this innovative technology to characterize the extent and volume of DNAPL in a shallow ground-water aquifer. Initial project tasks included hydrostratigraphic delineation using ground penetrating radar surveys, cone penetrometer tests, and drilling. Subsequently, a divergent–flow line drive wellfield was designed and hydrogeologically tested with slug tests and conservative interwell tracer tests. Laboratory work and UTCHEM simulations were also used to characterize the geosystem and design the PITTs. The tracer tests yielded response curves that were analyzed to determine the spatial distribution and total volume of DNAPL in the aquifer. The results provide a definitive estimate of the amount of DNAPL remaining in the DNAPL source at OU2, and demonstrate that it is practicable to determine the spatial distribution and volume of DNAPL in heterogeneous alluvium using partitioning tracers. This information provides remedial engineers with a design basis for enhanced DNAPL recovery operations and allows the prediction of accurate life-cycle clean-up costs.

INTRODUCTION

The disposal of chlorinated degreasing solvents in unlined trenches at Operable Unit 2 (OU2), Hill AFB, UT during the period 1967 to 1975 contaminated a shallow alluvial aquifer with multi-component dense non-aqueous phase liquid (DNAPL) composed primarily of trichloroethene (TCE), other chlorinated solvents, and oil and grease. In this case, the volume of DNAPL released was large relative to the aquifer pore volume available to trap the free-phase DNAPL. Hence, pooling of the DNAPL beneath the disposal trenches occurred within the alluvium above a clay aquiclude. Even though over 38,000 gallons (144,000 liters) of DNAPL have been recovered from the site, a large volume of residual DNAPL remains trapped by capillary forces in the pore spaces of the aquifer and continues to act as a source of dissolved contamination. Despite five years of continual ground-water extraction, pump-and-treat technology has failed to remove the source of contamination. This paper presents the results of the first full-scale field application of the partitioning interwell tracer test (PITT) to characterize a DNAPL source zone in terms of the spatial distribution and volume of DNAPL contained within it.

The motivation for characterizing the DNAPL source zone beneath OU2 arose from the need to prevent further off-site contamination, and to provide a design basis for the restoration of the site. Incomplete knowledge of the spatial distribution of DNAPL has resulted in a myriad of remediation failures, serving to emphasize the necessity of detailed characterization of DNAPL zones before remediation by whatever technology chosen. Therefore, the primary objective of the DNAPL source delineation effort was to undertake a detailed characterization of the DNAPL-contaminated aquifer by partitioning tracer testing.

The OU2 Site. OU2 is situated on a terrace overlooking the Weber Valley on the northeast boundary of the base. The site is located on a bench situated on an east-facing slope cut into deltaic sediments deposited into the ancestral Great Salt Lake by the Weber River. In 1996, characterization work conducted during two surfactant flood demonstrations at OU2 (Brown et al., 1997; DE&S, 1997; Rice et al., 1998) indicated that aquiclude depressions trapping DNAPL were part of a buried paleochannel that was deeply incised into the thick underlying clay. This buried alluvial channel is situated directly below the location of the former "chemical disposal pits". Because the nature and extent of the paleochannel and the DNAPL trapped within it were not fully known, it was decided to delineate the paleochannel and to quantify the volume and extent of the DNAPL contamination using PITTs (USAF, 1999).

The PITT. Studies of residual DNAPL distribution in heterogeneous, sandy aquifer materials indicate that sediment cores are unlikely to either locate or provide reliable estimates of the volume of DNAPL at the field scale. This is true because the representative elementary volume of residual DNAPL is very much larger than that provided by cores (Mayer and Miller 1992). The PITT was developed at the University of Texas at Austin (UT) during the early 1990s and was first used by the oil industry to measure the residual oil saturation that is homogeneously distributed in water-flooded well fields. The method was modified so that not only the residual oil saturation but also the swept pore volume, and therefore the total volume of heterogeneously distributed NAPL could be determined. Therefore, the PITT allows the detection and volume estimation at field scale of both DNAPL and light non-aqueous phase liquids (LNAPL). PITTs can be used in the saturated zone using ground water as the tracer carrier fluid, or in the vadose zone using air as the tracer carrier. The experimental and theoretical basis for the use of partitioning tracers is presented in detail in Jin (1995), Jin et al. (1995), and Dwarakanath (1997). The first environmental application of a field-scale PITT took place in 1994 (Annable et al., 1998), and approximately 25 pilot-scale PITTs have been conducted at various sites across the US before the first full-scale application of PITTs described here.

A PITT consists of the injection of a suite of conservative (non-partitioning) tracers and partitioning tracers into one or more wells and the subsequent production of the tracers from one or more nearby extraction wells. After the tracer slug is injected, potable water is injected to drive the tracers across the zone of interest. In the steady-state flow field into which a tracer slug has been introduced, the dissolution of the partitioning tracers into the DNAPL in the pore spaces proceeds until equilibrium partitioning has been achieved. Following the cessation of tracer injection, the partitioning tracer concentration in the ground water adjacent to the residual DNAPL will decline because of continuing injection of tracer-free water. Consequently, the net flux of partitioning tracers will be from the DNAPL back into the ground water to preserve the equilibrium partitioning dictated by the particular partition coefficient for the tracer. The dissolution into the DNAPL and the subsequent exsolution of the partitioning tracers back into the ground water retards these tracers relative to the conservative tracers unaffected by the presence of DNAPL. The observed temporal separation of the tracers at an extraction well due to the partitioning process is analogous to the chromatographic separation of peaks in the column of a gas chromatograph (GC).

CHARACTERIZING THE OU2 GEOSYSTEM

A successful PITT requires developing a conceptual model of the DNAPL zone, which is referred to as the "geosystem". The first 'layer' of the OU2 geosystem is a hydrostratigraphic map of the major geological features within and adjacent to the DNAPL source zone and their three-dimensional geometry. Additional 'layers' include the physical and chemical properties of the alluvial deposits containing the DNAPL and the thick clay formation underlying the DNAPL. DNAPL samples provide important information on the physical-chemical properties of the immiscible liquid with respect to its original migration and trapping and its potential for remobilizing during remediation. Ground water within the alluvium comprises another important 'layer' because it acts as the transporting agent for the dissolved-phase contamination. Finally, the PITT provides the spatial distribution and volume of the DNAPL within the source zone. The complete geosystem constitutes the essential information for designing a successful remedial operation to remove the DNAPL irrespective of the technology employed. Incorporating this information in a multi-phase, multi-component simulator results in a model referred to as a 'geosystem model.

Hydrostratigraphy. The first project task involved characterizing the paleochannel, and the alluvial aquifer contained within it, to the north and south of the previously known source area. This was accomplished using a number of techniques including ground-penetrating radar (GPR), cone penetrometer testing (CPT), and hollow-stem auger drilling and coring to confirm the depths of sedimentary interfaces recorded by the GPR survey and CPT logs. During this process additional free-phase DNAPL was discovered in the alluvium both inside the containment wall and outside and to the north of the containment wall. These efforts culminated in the completion of the map shown in Figure 1 of the top of the Alpine Formation (the clay) upon which is deposited the Provo Alluvium (the paleochannel topography).

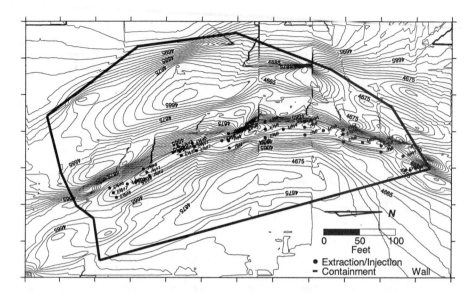

Figure 1. Topography of the Clay Aquiclude at OU2

Wellfield Design, Installation and Hydrogeologic Testing. Completion of these site characterization activities permitted the design of the well fields in which the PITTs were to be conducted. This design used an innovative and relatively inexpensive approach to conducting PITTs over large distances (110 ft) referred to as the divergent-flow line-drive well pattern. This pattern consists of a line-drive geometry for the well field in which the line of injection wells is positioned in the center of the well field and is flanked by two lines of extraction wells at opposite ends of the well field (see Figure 1). Thus, during injection tracers flow simultaneously in opposite directions towards the two lines of extraction wells. This innovation permits a particular volume of aquifer to be tested by partitioning tracers in half the time that would be required if the tracers were injected at one end of the well field and extracted at the other.

The next step was to install 27 new wells in four arrays in each of which a PITT would be conducted. These wells, in addition to 13 existing wells, comprised the four well arrays used during the PITTs. The new wells, and the alluvial-sediment cores collected during their installation, were tested to determine the hydraulic properties of the aquifer. Pneumatic slug tests and a conservative interwell tracer test (CITT) using NaCl as the tracer were conducted in each well array to provide design data for the PITTs, in particular, information on aquifer heterogeneities, swept pore volumes, and how each flow system responds to injection and extraction. Furthermore, each CITT mobilized and recovered additional remaining free-phase DNAPL, thus improving the accuracy of the subsequent PITTs.

DNAPL Properties. Cores from the newly completed wells were used in laboratory experiments to determine (1) appropriate partitioning tracers for use in the PITTs, (2) point estimates of DNAPL saturation using the partitioning theory code NAPLANAL (Mariner et al., 1997), and (3) the potential for mobilization of DNAPL by capillary desaturation. These latter experiments and subsequent Amott tests indicated that the DNAPL appears be wetting some fraction of the alluvium ("mixed-wet condition"). The mixed-wet condition of the aquifer sediments indicates that within portions of the alluvium the DNAPL prefers the smaller pore spaces, a condition that must be taken into account in selecting and designing a remediation technology. The DNAPL was determined to be primarily TCE (~60%), other chlorinated degreasing solvents (~15%), and oil and grease (~25%) with the following physical-chemical properties: density ~ 1.38 g/mL, viscosity ~ 0.78 cP, and interfacial tension ~ 9 dynes/cm. Given the origin of the DNAPL disposed at OU2 during 1967-75, this chemical composition and these physical-chemical parameters can be considered representative of typical chlorinated degreasing solvents.

PITT DESIGN AND IMPLEMENTATION

The information from the field and laboratory studies was incorporated into a revised geosystem model using the UTCHEM simulator (Delshad et al., 1996). This revised model, incorporating hydraulic information obtained from the CITTs, was then used to design PITTs for each well array. The average hydraulic conductivity of the Provo alluvium in the revised model well arrays was of the order range from 28 ft/day (1 x 10-4 m/s) in the northernmost well array (#1) to 71 ft/day (2.5 x 10-4 m/s) in well array #3.

The four PITTs were conducted from June through November 1998. The completion of these PITTs scanning a total swept pore volume of 285,000 gallons (>1 million liters) in four different well patterns was made possible by the use of innovative and mobile process systems. An electronic control system monitored and supervised the injection and extraction of fluids in the well array, and controlled sampling from various monitoring locations. It also logged all system parameters such as injection and extraction rates, the specific electrical conductivity of the effluent, and the various water pressures in the well array. Flow rates for each injection and extraction well were controlled by an automated flow-control system mounted in a trailer. The length of the PITTs ranged from 18 to 24 days, with a preliminary 2 to 4 day water-flood to achieve a steady-state flow field, a 1 to 2 day tracer injection period, followed by a water flood to transport the tracers through the aquifer.

Production of the tracers at the extraction well arrays yielded tracer response curves that were analyzed by the method of first moments to determine the spatial distribution and total volume of DNAPL in the aquifer. An example of such a curve is presented in Figure 2. Results are summarized for a range of aquifer porosity in Table 1. These values indicate that the majority of the remaining DNAPL in place is in the northern end of the aquifer in well arrays 1 and 2, which constitute the deepest part of the paleochannel. A visual

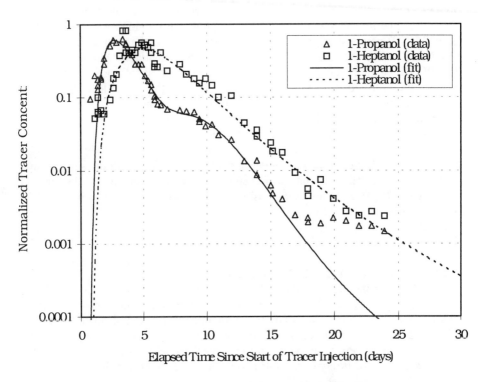

Figure 2. Example of PITT Breakthrough Curves

Table 1. Estimated Average DNAPL Saturations for the Four Well Arrays

Well Array (half)	Estimated Pore Volume below 4654 ft amsl (gal)		Volume of DNAPL (gal)	Estimated Average DNAPL Saturation(%)	
	27% Porosity	30% Porosity		27% Porosity	30% Porosity
1 north	1,135	1,261	122	10.7	9.7
1 south	5,597	6,219	192	3.4	3.1
1 (total)	6,732	7,480	314	4.7	4.2
2 north	10,189	11,321	340	3.3	3.0
2 south	13,661	15,178	321	2.3	2.1
2 (total)	23,849	26,499	661	2.8	2.5
3 north	14,065	15,628	51	0.4	0.3
3 south	6,103	6,782	27	0.4	0.4
3 (total)	20,169	22,410	78	0.4	0.3
4 north	5,000	5,555	36	0.7	0.6
4 south	846	940	44	5.2	4.7
4 (total)	5,846	6,496	80	1.4	1.2
Total	52,597	62,885	1,133	2.0	1.8

representation of the average DNAPL saturation distributed throughout the swept volume is presented in Figure 3. The plot also shows the surface of the Alpine Formation, and the containment wall encompassing most of the source zone. The average DNAPL saturation within the portion of the swept pore volume that is contaminated by DNAPL is estimated to be 2%, with a range from 11% at the north end of the source zone to less than 0.4% in the southern portion.

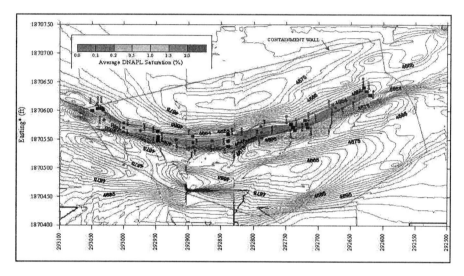

Figure 3. Average DNAPL Saturations in the Source Zone at OU2

CONCLUSION

Adding the results of the PITTs to the amount of DNAPL previously recovered from the source zone provides a new estimate of approximately 45,900 gallons (170,000 liters) of DNAPL originally residing within the pool encompassed by the containment wall. This result constitutes a quantitative characterization of a large DNAPL zone at the field scale. The data obtained during the course of this work provide a quantitative estimate of the volume and extent of the DNAPL source zone within the containment area of OU2. This information provides a remedial design basis; that is, all the necessary data for the evaluation and selection of source removal approaches. Finally, the completion of the source delineation at OU2 demonstrates that it is practicable to determine the spatial distribution and volume of DNAPL in heterogeneous alluvium using partitioning tracers.

REFERENCES

Brown, C. L., M. Delshad, V. Dwarakanath, R. E. Jackson, J. T. Londergan, H. W. Meinardus, D. C. McKinney, T. Oolman, G. A. Pope, and W. H. Wade. 1999. "Demonstration of Surfactant Flooding of an Alluvial Aquifer Contaminated with

DNAPL." In *Innovative Subsurface Remediation*, ACS Symposium Series #725, American Chemical Society, Washington DC.

Delshad, M., G. A. Pope, and K. Sepehrnoori. 1996. "A Compositional Simulator For Modeling Surfactant-Enhanced Aquifer Remediation: 1. Formulation." *Journal of Contaminant Hydrology*. 23: 303-327.

Duke Engineering and Services (DE&S). 1997. *Demonstration Of Surfactant-Enhanced Aquifer Remediation Of Chlorinated Solvent DNAPL At Operable Unit 2, Hill AFB, Utah.* Prepared for the Air Force Center for Environmental Excellence, Technology Transfer Division, Brooks Air Force Base, San Antonio, Texas, Draft Final.

Dwarakanath, V. 1997 "Characterization And Remediation Of Aquifers Contaminated By Nonaqueous Phase Liquids Using Partitioning Tracers And Surfactants." Ph.D. Dissertation, The University of Texas, Austin, TX.

Rice University, DE&S, University of Texas at Austin, and Radian. 1997. *Surfactant/Foam Aquifer Process of Remediation - Draft Final.* Prepared for The Advanced Applied Technology Demonstration Facility, Rice University, Houston, Texas.

Jin, M. 1995. "Surfactant Enhanced Remediation And Interwell Partitioning Tracer Test For Characterization Of NAPL Contaminated Aquifers." Ph.D. dissertation, University of Texas, Austin, TX.

Jin, M., M. Delshad, V. Dwarakanath, D. C. McKinney, G. A. Pope, K. Sepehrnoori, C. E. Tilburg, and R. E. Jackson. 1995. "Partitioning Tracer Test For Detection, Estimation And Remediation Performance Assessment Of Subsurface Nonaqueous Phase Liquids." *Water Resources Research*. 31(5): 1201-1211.

Mariner, P. E., M. Jin, and R. E. Jackson. 1997. "An Algorithm for the Estimation of NAPL Saturation and Composition from Typical Soil Chemical Analyses." *Ground Water Monitoring & Remediation*. 17(2): 122-129.

Mayer, A. S. and C. T. Miller. 1992. "The Influence Of Porous Media Characteristics And Measurement Scale On Pore-Scale Distributions Of Residual Nonaqueous Phase Liquids." *Journal of Contaminant Hydrology*. 11: 189-213.

United States Air Force (USAF). 1999. *Final Report: OU2 DNAPL Source Delineation Project.* Prepared for Hill Air Force Base, Utah by URS Greiner Woodward Clyde and Duke Engineering and Services.

TRICHLOROETHENE PLUME SOURCE AREA
DELINEATION IN THE PREAKNESS BASALT

Brian A. Blum (McLaren/Hart, Inc., Warren, NJ)
Gary M. Fisher (Lucent Technologies Inc., Murray Hill, NJ)

ABSTRACT: Cost effective and timely mapping of a trichloroethene (TCE) source area in a competent fractured bedrock regime located within the Newark Basin of New Jersey was successfully accomplished (after other traditional methods had failed) using conventional, construction-related rock drilling equipment. The drilling was performed using the conventional hydraulic push (Geoprobe™) method as well as an innovative application of a track-mounted pneumatic rock drilling technique typically used for advancing boreholes for blasting. Real-time analytical data were obtained from an on-site mobile laboratory. The source area mapping was conducted in an iterative fashion: data derived from a borehole were used to select the next drilling locations in order to focus in on the hot spot. The bulk of the TCE source area was located in the vicinity of a nearly vertical plunging fault that transects the site. The mapped source area was defined as that area having TCE concentrations of 1.0 milligram per kilogram (mg/kg) in soil (New Jersey Department of Environmental Protection [NJDEP] Impact to Groundwater Soil Cleanup Criterion), and 10 milligrams per liter (mg/L) in groundwater (approaching approximately 1 percent of the solubility of TCE in water).

INTRODUCTION

A source area investigation, consisting of a soil boring, rock boring and analytical program, was conducted in connection with a remedial investigation (RI) at a research and development facility in New Jersey (the site). The RI was conducted to determine the nature and extent of a plume of dissolved TCE and other associated chlorinated aliphatic hydrocarbons (CAHs) in groundwater in order to choose the most appropriate remedy. The specific objective of the investigation was to locate the source of the dissolved TCE plume in groundwater and determine its extent, as defined by concentrations of TCE in both soil and groundwater.

The investigation was difficult because of the relatively large extent of the plume and the lack of any historic records to direct the search for the source area. For these reasons and because the source was limited to a relatively small area within saprolite (weathered and decomposed bedrock) and very competent basalt bedrock, conventional remedial investigation methods, such as soil gas surveys, soil borings, and monitoring wells, had failed. The innovative application of a pneumatic rock coring machine (the type used to set up core holes for blasting preparation) allowed the investigation to proceed in a timely manner and resulted in the location and delineation of the source area in the groundwater within a competent basalt unit.

The source area was defined as having TCE concentrations of 1.0 mg/kg in soil (NJDEP Impact to Groundwater Soil Cleanup Criterion), and 10 mg/L in groundwater (approximately 1 percent of the solubility of TCE in water – the rule for defining free or residual product in New Jersey's *Technical Requirements for Site Remediation* [N.J.A.C. 7:26E]).

SITE DESCRIPTION AND PHYSICAL SETTING

The site lies within the Newark Basin in the Piedmont Physiographic Province of New Jersey. The Newark Basin is an elongate northeast-southwest trending fault trough filled with fluvial and lacustrine sediments of latest Triassic and earliest Jurassic age and three basalt flows of early Jurassic age. The site is situated on top of the second basalt flow (Preakness Basalt unit), also known as the Second Watchung Mountain, which exhibits a steep slope to the south and east while slopping more gradually to the north and west.

Land surface elevations at the site range from approximately 548 feet (167 meters) National Geodetic Vertical Datum (NGVD) at the southeast corner of the site (near the Building No. B) to 410 feet (125 meters) NGVD in the northern part of the site. A site map is provided in Figure 1.

Located in a suburban area, the facility has been home to research and development operations since the 1940s. It consists largely of laboratory and office space, comprising 2,000,000 square feet (185,800 square meters).

SITE GEOLOGY

Four significant stratigraphic units from the land surface downward underlie the site (Figure 2): glacial till, saprolite, the Preakness Basalt, and the Feltville Formation (sedimentary rock). With a few exceptions, these stratigraphic units do not yield significant quantities of water to discrete interval monitoring wells. The till and saprolite are dense and the bedrock is very competent.

New Jersey Geologic Survey Fault Index No. 147 outcrops at the cliff-face on an interstate highway and passes through the source area as it transects the site. New Jersey Geologic Survey Fault Index No. 145 transects the site just west of the source area. These two faults that exhibit a near vertical plunge (85°) were observed to trend into the site in the vicinity of Building No. B (Figure 3). The faults have been mapped by the New Jersey Geological Survey (Drake, A. A., *et al.* 1996) as intersecting the Feltville Formation.

HYDROGEOLOGY

Site monitoring wells (Figure 1) are screened across four separate zones. Wells screened in the "shallow" zone are within the unconsolidated deposits (till and saprolite). Wells in the "intermediate" zone are screened in the upper Preakness Basalt, and wells in the "deep" zone are screened at least 100 feet (30.48 meters) into the Preakness Basalt. Wells screened in the "sedimentary"

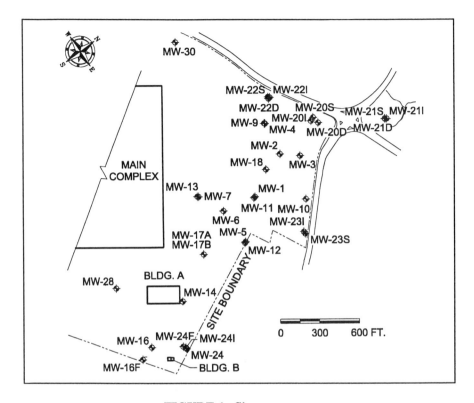

FIGURE 1. Site map

zone are within the Feltville Formation. Groundwater in the "shallow", "intermediate", and "deep" zones has a horizontal potential of flow to the north to northeast. The horizontal component of flow in the "sedimentary" zone is toward the west.

Faults that transect the site play a significant role in the migration pattern of the TCE plume in groundwater. The faults limit the horizontal migration of the plume in the Preakness Basalt and likely influence the vertical migration of contaminants to the Feltville Formation. The water-bearing deposits underlying the site are primarily low yielding and exhibit relatively low hydraulic conductivity.

HISTORIC INVESTIGATIONS

In 1982, a groundwater monitoring installation and sampling program was initiated at the site as part of a New Jersey Pollutant Discharge Elimination System discharge to groundwater permit requirement for a holding pond. Monitoring well sampling results indicated the presence of CAHs (predominantly TCE) at concentrations above NJDEP Groundwater Quality Standards. Since 1982, several groundwater and soil investigation programs aimed at defining the

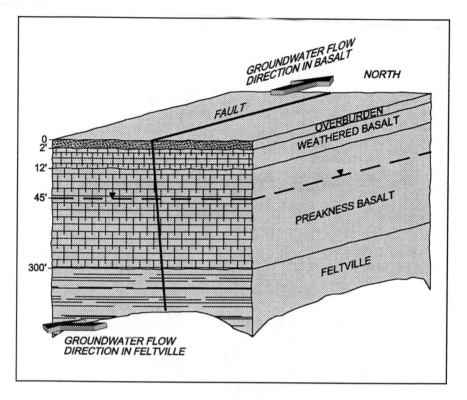

FIGURE 2. Schematic Geologic Cross Section

nature and extent of the groundwater plume and identifying the source were
conducted.

Numerous investigations were successful in identifying the plume and
developing a conceptual model of the physical and geologic setting of the site;
however, they did not completely delineate the plume or identify its source. None
of the "conventional" RI techniques, described below, were successful in
identifying the source of the TCE plume, primarily because there were no records
to indicate where TCE would have been released to the environment. In addition,
the source is a relatively small area where the overburden materials are comprised
of saprolite.

Non-Passive Soil Gas. In 1987, a soil gas study was conducted hydraulically
upgradient of the initial monitoring well network that defined the volatile organic
compound (VOC) plume. Soil gas was extracted at 18 locations, using a battery-
operated peristaltic pump, through a slotted probe that had been driven 2 to 3 feet
(0.61 to 0.91 meters) below land surface. The pump discharged the soil gas
directly into a gas chromatograph (GC) for analysis. Only six of the eighteen
samples exhibited detections of TCE above the instrument detection level.

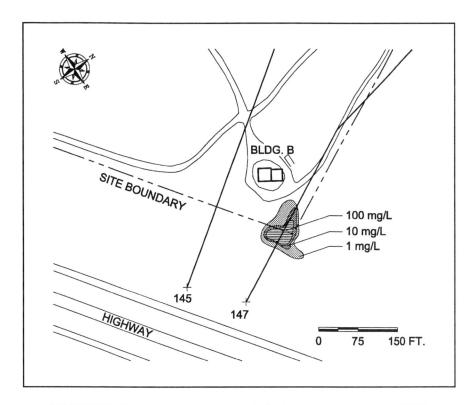

FIGURE 3. Source area with geologic faults and groundwater TCE isoconcentrations.

Soil Borings. Based on the results of the soil gas survey conducted in 1987, a soil-boring program was initiated. The program consisted of 34 soil borings. The soil samples were collected with conventional split barrel cores (split spoons). The borings were laid out in a generalized grid pattern between Building No. A and the eastern property line (Figure 1). Soil was field-screened using a Photovac portable GC. Samples were collected at 2 to 5 foot (0.61 to 1.52 meter) intervals until either bedrock or the water table was encountered. Where water was encountered, a sample was taken and a head space analysis was performed. The program identified low concentrations of TCE in the soils east and southeast of Building No. A and concluded that TCE found in site monitoring wells may have originated in the vicinity of Building No. A.

Well Installation. In 1990, a monitoring well installation program was conducted to further delineate the TCE plume and determine the source area. The program consisted of the installation of seven monitoring wells screened in the shallow bedrock (saprolite) (Figure 1). This well installation program concluded that the area east of Building No. A was believed to be the source of TCE contamination and that the current source of contamination is residual TCE remaining in the fractures of the bedrock.

RECENT INVESTIGATION

A source area investigation was initially focused in the area east of Building No. A because the highest concentrations of TCE in groundwater were observed in monitoring Well MW-24 (Figure 1) which had been installed as part of the site RI. The TCE plume that extended throughout the eastern part of the site (Figure 1) was presumed, because of its shape and topographic control, to originate in the southeast part of the site, near Building No. A. Only a relatively small area exists between the presumed plume origin and an 80-foot (24.4-meter) drop-off toward an interstate road cut. A pilot scale experiment with a passive soil gas technique in the vicinity of MW-14 proved to be ineffective for detecting TCE vapors from the groundwater; therefore, it was determined that an aggressive technique for sampling shallow groundwater, within the competent basalt, was warranted to find the source.

Conventional rock coring or drilling to the first water was inappropriate due to the relatively high cost per linear foot (meter), its relatively slow speed, and the access constraints of most rotary pneumatic drilling equipment. An acceptable alternative was to use the widely available, versatile equipment that is made specifically for rapidly advancing boreholes into rock. This track mounted pneumatic rock corer (an Ingersol Rand ECM-590 Hydraulic Drill [see Figure 4]) is normally used to drill core holes for blasting.

The rock borings initially proposed were along a grid extending from MW-24 toward Building No. B. However, since initial borings in the vicinity of MW-24 and in the corners of the proposed grid did not indicate any source-related TCE concentrations, the approach was modified to locations outside the original grid and was based on analytical findings of high TCE concentrations. A total of forty-five rock borings (32 in Phase 1 and 13 in Phase II) were drilled vertically and three borings were drilled at angles ranging from $10°$ to $15°$ to intercept one of the faults that transects the site (Figure 3).

During the initial phase (Phase 1) of the source area investigation, groundwater was encountered at 25 of the 35 drilling locations and groundwater samples were collected with dedicated Teflon® bailers. A New Jersey licensed laboratory analyzed all samples for VOCs by US EPA Method 624 and provided analytical results within 24 to 48-hours of sample pick-up. Phase 1 was successful in locating the source area, but in order to define the boundaries, a more rapid analytical turnaround ("real time" data) was needed.

In Phase II of the source area investigation a New Jersey-certified mobile laboratory was brought to the site. It further expedited the turn-around-time required for the groundwater samples and subsequent soil samples, and thereby allowed an integrated "real time" approach to the delineation program. During this second phase of the source area investigation, 12 groundwater samples and 40 soil samples were analyzed within a four-day period. The groundwater sampling and analysis were conducted concurrent with the soil sampling. Locations with

FIGURE 4. Photograph of an Ingersol Rand ECM-590 Hydraulic Drill Rig

elevated groundwater contamination invariably indicated areas that contained saprolitic soils with TCE. Soil sampling was conducted by the Geoprobe™ method. Table 1 below summarizes the rock drilling and groundwater sampling activity during the source area investigation.

TABLE 1. Source Area Investigation Rock Drilling Summary

Investigation Phase	No. of Borings	No. of Borings with Groundwater	Total Feet (Meters) Drilled
Phase I	35	25	2,088 (636.4)
Phase II	13	12	793 (241.7)

TECHNIQUE LIMITATIONS

The maximum drilling depth of the equipment used was 80 feet (24.4 meters) deep; however, more powerful models are capable of drilling deeper. Since the rock boreholes are not cased off, a large overburden depth may result in borehole cave-ins. The ECM-590 does not have a sufficient air supply to blow out significant amounts of debris from the borehole. Also, since a relatively small diameter open borehole is drilled, the technique is not useful for advancing casing or packers to allow for vertical profiling.

RESULTS AND CONCLUSIONS

Investigation activities at the site identified a TCE plume extending over 2,500 feet (762 meters) and to a depth of 600 (183 meters) feet within till, saprolite, and two competent bedrock formations. Dissolved TCE concentrations in the plume range from less than 10 micrograms per liter ($\mu g/L$) to greater than 5,000 $\mu g/L$. The source area of the TCE was determined to be in the southeast corner of the site, overlying a significant geologic fault. Soils (weathered rock) with TCE concentrations above the soil criterion of 1 mg/kg extend over an area of approximately 600 sq. ft. (55.7 square meters) (15 ft [4.6 m] x 40 ft [12.2 m]). Groundwater with TCE concentrations above the solubility criterion of 10 mg/L covers approximately 4,800 sq. ft (445.9 square meters) (60 ft [18.3 m] x 80 ft [24.4 m]) (see Figure 3).

The inherent difficulties of site characterization and remediation due to the presence of DNAPL made determining the location and extent of the source zone critical. In this setting, the application of the pneumatic rock coring equipment was key to identifying and determining the extent of the TCE source area. With its mobility, quick application, and lower costs, the pneumatic rock drilling technique is ideal for delineating contaminants that were introduced directly into competent bedrock with little or no overburden.

REFERENCES

Drake, Avery A., et al. 1996. *Bedrock Geologic Map of Northern New Jersey*. Miscellaneous Investigations Series Map I-2540-A, scale 1:100,000. U.S. Department of Interior.

Technical Requirements For Site Remediation (Adopted July 1, 1997). New Jersey Administrative Code (N.J.A.C.) 7:26E.

DNAPL CHARACTERIZATION USING THE RIBBON NAPL SAMPLER: METHODS AND RESULTS

Brian Riha, Joe Rossabi, Carol Eddy-Dilek, and Dennis Jackson (Westinghouse Savannah River Company, Aiken, SC), Carl Keller (Flexible Liner Underground Technologies, Ltd., Santa Fe, NM)

ABSTRACT: The Ribbon NAPL Sampler (RNS) is a direct sampling device that provides detailed depth discrete mapping of Non Aqueous Phase Liquids (NAPLs) in a borehole. This characterization method provides a yes or no answer to the presence of NAPLs and is used to complement and enhance other characterization techniques. Several cone penetrometer deployment methods are in use and methods for other drilling techniques are under development. The RNS has been deployed in the vadose and saturated zones at four different sites. Three of the sites contain DNAPLs from cleaning and degreasing operations and the fourth site contains creosote from a wood preserving plant. A brief description of the process history and geology is provided for each site. Where available, lithology and contaminant concentration information is provided and discussed in context with the RNS results.

INTRODUCTION

Dense Non Aqueous Phase Liquids (DNAPLs) such as solvents and dry cleaning fluids migrate downward and are often present in the subsurface as small discrete globules or lenses which are difficult to locate using traditional characterization methods. NAPLs arc a common long-term groundwater contamination source. If the long-term source can be found, more aggressive remediation efforts can be used to clean the source and reduce the long-term impact on the aquifer and the associated costs of treating large dissolved contamination plumes.

The Ribbon NAPL Sampler (RNS) provides continuous depth discrete sampling in a borehole and immediate analysis results for the pure phase component. This characterization method determines the presence or absence of NAPLs and is used to complement and enhance other characterization techniques. The RNS also works for Light Non Aqueous Phase Liquids (LNAPLs). The Ribbon NAPL Sampler and installation methods are discussed as well as characterization results from four different sites.

MATERIALS AND METHODS

The Ribbon NAPL Sampler is a direct sampling device that provides detailed depth discrete mapping of NAPLS in a borehole. This characterization technique uses the Flexible Liner Underground Technologies, Ltd. (FLUTe) membrane system (patent pending) to deploy a hydrophobic absorbent ribbon in the subsurface. The system is pressurized against the wall of the borehole and the ribbon absorbs the NAPL that is in contact with it. A schematic of the RNS is shown in Figure 1.

The FLUTe membrane consists of an airtight liner that is pneumatically and/or hydraulically installed in a borehole. The rugged flexible tubular membrane supports and seals the borehole wall and can be installed in the saturated and vadose zones by several techniques. The membrane technology has been used to place sampling ports and sensors in varying sized boreholes to depths of 800 ft. Removal of the membrane is accomplished by turning the membrane inside out by pulling on a tether connected at the bottom of the liner. The membrane can be reused for multiple deployments.

The absorbent ribbon is a sleeve that covers the FLUTe membrane and is manufactured from a material that will repel water and absorb liquid solvents and petroleum products (NAPLs). This hydrophobic material readily "wicks" NAPL compounds from the adjacent borehole sediments. The primary analysis method uses a hydrophobic ribbon impregnated with a powdered oil dye (Sudan IV). The dye dissolves in NAPLs that are absorbed into the ribbon and stains the ribbon bright red. The ribbon is replaceable for additional deployments with the same FLUTe membrane. A characteristic spot from residual DNAPL detection with the RNS is shown in Figure 2.

In non-collapsing vadose zone boreholes, the Ribbon NAPL Sampler is deployed with air pressure. The hydrophobic ribbon is attached to

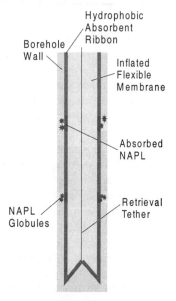

Figure 1. RNS schematic

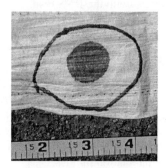

Figure 2. Characteristic spot from DNAPL detection with the RNS

the membrane and the membrane is everted (turned inside out) from a pressure canister. This eversion method prevents the ribbon from sliding along the borehole and smearing the NAPL on the membrane. The membrane is retrieved and then re-everted at the surface and inspected for the presence of NAPL. The reusable membrane is available in custom lengths and can use any length of the replaceable hydrophobic ribbon. A two-inch diameter membrane is used in CPT boreholes and other diameters are available.

The installation method for the Cone Penetrometer (CPT) allows for installing the Ribbon NAPL Sampler below the water table and in collapsing sediments in the vadose zone. The RNS is fabricated with a bundled ribbon around the membrane and comes assembled to specified lengths from FLUTe Ltd.

One of the current designs is for the standard CPT rods with a 1.75 inch OD and 1 inch ID. Once the CPT rods are pushed to depth, the bundled RNS is lowered into the CPT rods and the rods are retrieved a few feet to release the sacrificial tip and anchor the membrane in the sediments. For each CPT rod retrieved, water is measured into the bottom inside of the membrane through the tether tube to expand the membrane and hold the borehole open. Water is also added between the membrane and CPT rods to balance the fluid pressure and reduce friction. Once all the rods are retrieved and the membrane has been in contact with the formation, the RNS is retrieved by pulling the tether up and turning the membrane inside out. The inversion brings the ribbon up on the inside away from the sediments. The water inside the RNS is clean. The RNS is turned right side out and the locations of depth discrete NAPL, indicated by dyed portions of the membrane, are recorded. The RNS can be rebuilt with a new bundled ribbon.

RESULTS AND DISCUSSION

The Ribbon NAPL Sampler has been deployed at four different sites in the vadose and saturated zones. Three of the sites contain DNAPLs from cleaning and degreasing operations and the fourth site contains creosote from a wood preserving plant. A brief description of the process history and geology is provided for each site. Where available, lithology and contaminant concentration information is provided and discussed in context with the RNS results.

SRS A-14 Outfall. The A-14 Outfall is located in the northern section of the DOE Savannah River Site (SRS). This area housed reactor fuel and target assembly fabrication facilities and laboratory and support facilities whose operations resulted in releases of chlorinated solvents. Beginning in 1952, the process wastes were released through a process sewer system to the A-14 outfall. Between 1952 and 1979, historical records estimate 1,395,000 lbs of solvents were released to the outfall. Of this quantity 72% was tetrachloroethylene (PCE) and 27% was trichloroethylene (TCE). Solvent discharge to the outfall ended in 1979. The outfall is currently used to discharge treated process water and storm water (Jackson et al., 1999). The geologic framework consists of heterogeneous interbedded layers of sand, silt and clay.

Four Ribbon NAPL Samplers were deployed in the vadose zone in open cone penetrometer boreholes at the SRS A-14 Outfall. These deployments were designated HFM-1, HFM-2, HFM-3, and MVE-17. Each of the samplers remained in contact with the borehole for approximately one hour. HFM-1 was installed approximately 20 ft away from the head of the outfall and did not indicate the presence of DNAPL. HFM-2 was deployed to a depth of 60 ft and is above the current outfall discharge. HFM-3 and MVE-17, deployed to a depth of 30 ft, are approximately 15 ft away from HFM-2, within a few feet of each other, and near the headwaters of the current outfall.

CPT information and DNAPL location identified by the RNS for HFM-2, HFM-3, and MVE-17 are presented in Figure 3. The heavier line is the CPT friction ratio where higher values indicate finer grained materials. The lighter line is the sediment resistivity where lower values correspond to more conductive sediments (i.e. wet and/or silty and clayey materials) and the higher values correspond to less conductive sediments (i.e. dry and/or sandy materials). The horizontal lines are the depth discrete location of DNAPL found with the RNS. Note the significant differences in geology between the boreholes and that HFM-2 is dryer than HFM-3 and MVE-17.

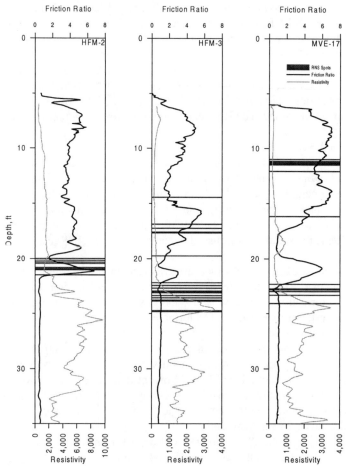

Figure 3. RNS and CPT Results from the SRS A-14 Outfall

DNAPL was found in HFM-2 in the fine-grained materials at approximately 20 ft. DNAPL is present in HFM-3 and MVE-17 in the 22.5-25 ft range which corresponds to the coarser material below the 20 ft fine grained material observed in HFM-2. This inconsistency can possibly be attributed to the constant water infiltration from the outfall moving the DNAPL downward and/or the DNAPL is not coming into contact with the hydrophobic ribbon due to water saturation in the clay. More DNAPL is observed deeper in HFM-3 than in MVE-17 which can be attributed to more coarse grained materials in HFM-3. DNAPL is observed at 12 ft in MVE-17 where downward migration is retarded in the fine-grained material. The nature of the staining on the ribbon indicates the DNAPL is in the form of dispersed globules or very thin lenses and is not present in pools or strong discrete layers.

Cape Canaveral Air Station Launch Complex 34. Launch Complex 34 (LC34), was constructed during 1959-1960 for the Saturn I and IB rockets, which served as launch vehicles during the early Apollo manned space program. The Saturn rocket engines were cleaned while on the launch pad with solvents, predominantly trichloroethylene (TCE), and the solvent was discharged to a nearby drainage pit. Engine parts were also cleaned with TCE at a nearby support building. These on-site cleaning activities ceased in 1968 with the termination of operations at the complex. The geologic framework consists of an upper unit of relatively uniform fine grained sand 23-27 ft thick from grade, a middle unit of silty and clayey fine sand 1-17 ft thick, a lower sand unit of silty fine sand 15 ft thick, and a confining clay unit at approximately 45 ft below grade. The water table was about 4 ft below surface during the deployment.

The Ribbon NAPL Sampler was deployed in the saturated zone through the cone penetrometer rods to a depth of 35 ft below ground surface. The deployment of the sampler took approximately 1.5 hours and it remained in contact with the borehole for approximately 45 minutes. The staining on the sampler showed residual DNAPL (TCE) is present from approximately 18 ft to 35 ft with a higher density sitting on top of the middle fine unit and at the bottom the middle fine unit. The spotty nature of the staining on the ribbon indicates the DNAPL is in the form of dispersed globules and is not present in pools or strong discrete layers. The location of DNAPL found with the RNS is shown in Figure 4 along with TCE sediment concentrations from a nearby boring completed during previous characterization work. At this particular site, DNAPL is present when sediment concentrations are between 250 and 450 mg/kg (Eddy-Dilek et al., 1999). The sediment sampling was conducted at 1 ft intervals to a depth of 48 ft.

Paducah Gaseous Diffusion Plant. The TCE leak site is located near a building used for degreasing, cleaning, and testing of components used in the gaseous diffusion process. One potential release at this site is from effluent leaking from a subsurface pipe carrying discharge from a sump in the building to a storm sewer. It was not known that the sump discharged to the storm sewer and the release may have begun as early as the 1950s. A pump station used to offload TCE from tank cars into a storage tank had leaked and been repaired several times in the past and released an unknown but significant quantity of TCE to the subsurface. The geologic framework consists of an upper 60 ft section of heterogeneous interbedded layers of clay, silt, sand, and gravel (CH2M Hill, Inc., 1999). The water table was located at about 40 ft below the surface.

The Ribbon NAPL Sampler was deployed through the cone penetrometer rods to a depth of 59 ft below ground surface. The deployment of the sampler took approximately 2 hours and it remained in contact with the borehole for approximately 45 minutes. The staining on the sampler showed residual DNAPL (TCE) is present throughout the sampled interval with the highest densities located at approximately 10 ft, 25-40 ft, and 50-55 ft. The spotty nature of the staining on the ribbon indicates the DNAPL is in the form of dispersed globules and is not present in pools or strong discrete layers. The location of DNAPL found with the RNS is shown in Figure 5 along with TCE sediment concentrations from a nearby boring completed during previous characterization work (CH2M Hill, Inc., 1999). The sediment sampling was conducted at 4 ft intervals to a depth of 29 ft. Unlike conventional sediment sampling and analysis, the Ribbon NAPL Sampler provides continuous sampling in a borehole with immediate field results.

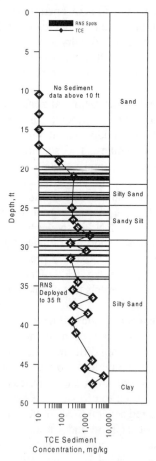

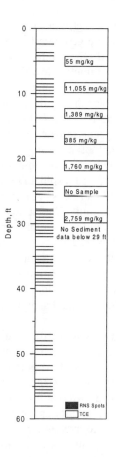

Figure 4. RNS results and sediment concentration from LC34

Figure 5. RNS results and sediment concentration from Paducah

EPA Superfund Site, Stockton, California. McCormick & Baxter Creosoting Co. formerly operated a wood-preserving facility on a 32-acre site in a light industrial area near the Port of Stockton. From 1942 to 1990, utility poles and railroad ties were treated with creosote, pentachlorophenol (PCP), and arsenic compounds. Waste oils generated from the wood-treatment processes were disposed of in unlined ponds and concrete tanks on-site. Previous characterization found soils throughout the site were contaminated with constituents of creosote. The geologic framework consists of heterogeneous interbedded gravel, sand, silt, and clay. The water table is located approximately 10-15 ft below surface (U.S. Army Corps of Engineers, Seattle District, 1999).

Two Ribbon NAPL Samplers were deployed through the vadose zone into the saturated zone through the CPT rods. Each of the samplers remained in contact with the borehole for approximately one hour and was deployed near previous CPT and laser induced fluorescence (LIF) characterization pushes. Location SE-05 was deployed to a depth of 39 ft and SE-39 was deployed to 44 ft. CPT and LIF information and DNAPL location identified by the RNS are presented in Figure 6. The CPT friction ratio is a measure of sediment types where higher values indicate finer grained materials. The LIF counts indicate the location of the creosote contaminants. At creosote contaminated sites, the creosote stains the ribbon a dark black or brown and overshadows the red dye staining.

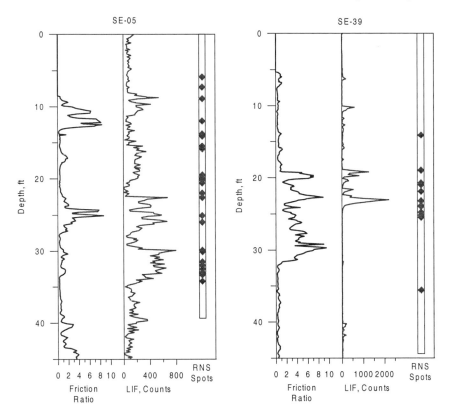

Figure 6. RNS results, lithology and LIF information from and sediment concentration from EPA Superfund Site, Stockton, California

In SE-05 the staining on the sampler showed DNAPL (creosote) is present throughout the sampled interval. The location of DNAPL found with the RNS correlates with some of the LIF peaks especially in the 30-35 ft range. The discrepancy can be attributed to the heterogeneity (retarded downward movement caused by the fine material at 10 and 25 ft) and the resulting dispersed, residual nature of the DNAPL in this area. The RNS was deployed within approximately 5 ft of the CPT push. In SE-39, the DNAPL found with the RNS correlates well with the LIF data. At this location, the DNAPL is primarily located in the upper portion of the 10 ft thick interbedded sand, silt, and clay zone starting at 20 ft.

CONCLUSIONS

The Ribbon NAPL Sampler has proven to be a robust method for determining the depth discrete location of DNAPL in the subsurface. The sampler is easily deployed with CPT in both the vadose and saturated zones. Deployment methods for both larger and smaller diameter drilling techniques are under development. The results from four DNAPL sites show DNAPL is present as dispersed globules and does not exist in pools or layers and is strongly controlled by lithology. Comparative results of sediment sampling and the RNS prove soil sampling is prone to missing the small dispersed globules of DNAPL whereas the RNS provides continuous sampling in a borehole. The RNS provides a complementary characterization technique for NAPL field screening and for verification of the presence and location of NAPL.

REFERENCES

CH2M Hill, Inc. 1999. *Remedial Investigation Report for Waste Area Grouping 6 at Paducah Gaseous Diffusion Plant, Paducah Kentucky.* DOE/OR/07-1727/V1&D2.

Eddy-Dilek, C. A., B. D. Riha, D. Jackson, J. Rossabi, J. Consort. 1998. *DNAPL Source Zone Characterization of Launch Complex 34, Cape Canaveral Air Station, Florida.* Westinghouse Savannah River Company Report, WSRC-TR-99-00024.

Jackson, D. G., W. K. Hyde, J. Rossabi, B. D. Riha. 1999. *Characterization Activities to Determine the Extent of DNAPL in the Vadose Zone at the A-014 Outfall of the A/M Area (U).* Westinghouse Savannah River Company Report, WSRC-RP-99-00569.

U.S. Army Corps of Engineers, Seattle District. 1999. *Management Plan for NAPL Field Exploration. McCormick and Baxter Superfund Site, Stockton, California.* EPA Region 9 Report.

DNAPL SITE CHARACTERIZATION: THE EVOLVING CONCEPTUAL MODEL AND TOOLBOX APPROACH

Rossabi, J., B. B. Looney, C. A. Eddy-Dilek, B. D. Riha, and D. G. Jackson, Westinghouse Savannah River Company, Aiken, SC 29808

Abstract: In natural subsurface systems dominated by heterogeneity, the delineation and even the detection of sparingly soluble, dense contaminants can be extremely difficult. The performance assessment of cleanup at these sites is therefore more complex. Several technologies for the characterization of sites contaminated with dense non aqueous phase liquids (DNAPLs) have recently been developed. These include geophysical techniques, tracer tests, and direct sampling or sensing methods. The innovative methods provide some significant advances over conventional sampling-based approaches but the real value of these methods is in their addition to a toolbox approach to DNAPL characterization. The toolbox approach recognizes that all characterization methods contribute to the conceptual model of the site. The strategic selection of technology and results from each application must contribute to the evolution of this conceptual model. The ultimate goal is the absolute knowledge of the contamination at the site. This, of course, can never be reached but using the right suite of tools and comprehensive integration of the data, the most accurate understanding is obtained.

INTRODUCTION

The evolving conceptual model and toolbox approach is simply a framework for the ideal understanding of a site. It can be used for the characterization of any site, not just DNAPL suspect sites. The principles involved are generally just good scientific and engineering practice. Unfortunately these principles are often overlooked in the face of complex systems or external pressures. Generally characterization methods should progress from less invasive to more invasive methods and from less expensive to more expensive. Simple, long-term observations should also not be overlooked (e.g. water level fluctuations in existing wells) and can often provide inexpensive and important keys to understanding a site. Finally it is useful to borrow from the experience of the geochemical exploration professionals – lots of inexpensive measurements (with slightly larger error bands) and fewer expensive (usually due to excessive quality assurance requirements) measurements. Of course, these are general guidelines and should not prevent the use of a very expensive but very useful method at the beginning of an investigation if it is justified for a particular site.

Despite the complexity of the situation or the encroaching time and economic pressures, the current conceptual model of the site must be logical, explainable, and defensible. A useful intellectual tool to isolate pressures that may cloud prudent decisions is to personalize the perspective - what would I do if this were my property? This mode of thinking maximizes the responsibility of the investigator because they are now thinking of a site in reference to scenarios where the protection of their family and their interests is involved. It also maximizes the attention to practical details such as cost and long term goals (e.g., remediation method and extent) for the site. The available technology is usually important in selecting characterization methods but the technology must be selected in the context of the remediation goals and requirements. If your planned remediation is vapor extraction you do not need to know the temperature gradient in the soil, at least initially.

DNAPL contaminated sites tend to be more complex than light non aqueous phase liquid (LNAPL) or aqueous contaminated sites because the physical and chemical characteristics of dense, sparingly soluble contaminants add additional complexity to heterogeneous geology and hydrogeology at most sites. While LNAPL contamination is usually constrained to the top of the water table and above, and aqueous phase contamination follows the hydrology of the site, DNAPL movement is controlled by gravity and capillary pressure of sediments, and can move against the hydraulic gradient.

The conceptual model is the snapshot or realization of the site at a particular time. It is optimally constructed through historical process information, the complete set of physical, chemical, geologic, and hydrogeologic data and understanding of the dynamic processes, experiential knowledge of similar sites, the risk scenarios related to the site and its contaminants, and the type and scope of remediation envisioned. The model is updated and refined by subsequent data and information and the process is iterative.

HISTORICAL INFORMATION, PROCESS HISTORY AND SITE COURTESY

At the Savannah River Site (SRS), we have used disposal records (amounts, types, timing), construction diagrams of drains, sewers, and basins (e.g., unlined manholes every 50 yards), aerial photos regional geologic and weather information, and we talked to the personnel who worked in the operations and were responsible for it. All of this information is not available at every site but much more is available then is generally used. A very important resource is the verbal record of site personnel. This information may not always be precise but it can be used with other information or to open an investigation path. Often the best information is obtained by allowing the personnel to explain what they did, saw, smelled, and tasted in their own narrative account with few directed questions that may influence the testimony. Understanding the daily operations at a site is also critical. How often were degreasing units used? How often were solvents purchased and in what quantities? How much process water was released per day? How often were machines serviced and what happened to the used lubricating oils? Were acids and caustics also released? Each of these questions can dramatically impact the deposition of contaminants at a site.

At SRS we have had the pleasure of hosting many different researchers and vendors with innovative technologies. We have found them to be nearly universally sincere in their desire to provide real solutions rather than carelessly promote themselves or their technology. One thing that we have experienced rarely, though often enough to mention, is the visitor to the site who believes he understands the host's problems better than the host does. A visitor may have some useful insight and ideas based on their experience in similar situations but the host knows his site best. We have learned this lesson and apply it when we visit other sites to help with their environmental problems.

BASELINE METHODS

DNAPL sites are rarely sought, they are generally stumbled upon based on unusually high concentrations found in the analytical results from a nearby aqueous or gas sample. The first suspicion of DNAPL arises from rules of thumb (e.g., 1 or 10% of aqueous solubility or vapor pressure) applied to these samples that are contaminated via dissolution and/or diffusion from a source. The value of the data from these routine samples should not be dismissed. Besides implying the presence of DNAPL, the diffusion-based samples can help focus more direct searches for DNAPL by concentration differences between aqueous or gas phase samples.

Baseline and modified baseline soil sampling is still the most reliable method for directly detecting DNAPL (Eddy Dilek et al., 1999). Using equilibrium partitioning rules, the presence of DNAPL in a soil sample can be determined (Cohen and Mercer, 1992). Some important modifications that should be incorporated include higher depth resolution in soil sampling. Because residual DNAPL at sites is often in small, dispersed blobs correlated to the soil type, the ideal depth resolution of sampling should be comparable to the resolution of the geologic heterogeneity. Traditional sampling generally prescribes a regular grid with samples collected at intervals of 5 feet at best. We have found that collecting samples every 1 to 2 feet and at lithologic changes has been significantly better than rigid 5-foot intervals. The cost of analysis for these samples can be defrayed by using less costly analytical techniques such as field screening methods or even EPA SW 846 methods (like heated headspace method 5021) without rigidly applying regulatory protocol or incurring the EPA Certified Lab Program paperwork. Of course, the local regulators should be consulted when pursuing this strategy. When continuously coring a drilled hole, collecting additional samples is very inexpensive. If the cost of analysis is prohibitive, a subset of the samples need only be analyzed. The rest can be cold stored for analysis later if desired. The literature has shown that VOC samples are stable for longer than the regulatory limit of 14 days, especially when the samples contain very high concentrations or DNAPL(West et al., 1996). If possible, aqueous samples should also be taken at higher vertical resolution by using depth-discrete samplers or setting wells with smaller screen zones. Heterogeneity and the ability to access and analyze at the scale of the heterogeneity limit subsurface characterization. The most common access techniques (drilling and direct penetration) have inherent lateral limitations but fewer vertical limitations. We should take better advantage of the accessibility to finer vertical resolution.

Despite the availability of more sophisticated methods of characterization, the ultimate performance assessment in the remediation of a site relies on the continued clean analytical results from down-gradient wells.

NONINVASIVE

Earlier we recommended the general strategy of progressing from non- or minimally invasive to more invasive. The most obvious choice for non-invasive methods beyond evaluating historical records and using existing infrastructure are the geophysical techniques. Acoustic and electromagnetic probing of the subsurface have been used successfully to detect reservoirs of petroleum or natural gas, or on a smaller scale, buried drums and other anomalous objects. They are also useful for detecting changes at a site through time (Daily et al., 1992). They probe volumes rather then measuring at discrete points, which can be an advantage in characterizing a site with small, dispersed pockets of contaminants. An important limitation of these methods, however, is that the resolving power is generally too low to detect the disparate blobs of DNAPL due to the inherent noise caused by subsurface heterogeneity. If pores were predominately NAPL filled and were contiguous over a large enough volume (usually on the order of a cubic meter), some geophysical techniques could directly detect DNAPL. A far more important use of the geophysical measurements is to characterize and differentiate the geologic units that would be likely to harbor DNAPL or control its movement. Like the lower resolution, diffusion based measurements, geophysics is very important for telling you where to look. The information provided from geophysics has even greater value when combined with other data from the conceptual model. At SRS, surface seismic and electromagnetic measurements were integrated to laterally map confining zones and determine their thickness.

In addition to surface geophysics, existing boreholes can often be used to enhance the resolution of surface techniques. While evaluating the conductivity logs at a known DNAPL site, anomalous patterns were evident in some of the wells near the release point (Nelson, 1996). Large quantities of caustics and acids were disposed with process water as a dense aqueous phase liquid (DAPL) at the same location as the solvents. The caustic solution presented a sharp electrical conductivity contrast to the normally low conductivity formation water at the site. The conductivity anomalies coincided with higher concentrations of organics indicating that the DAPL may have traveled in the same path as the DNAPL until dilution of the caustic solution rendered it neutrally buoyant. These boreholes may also be suitable for long term measurements of the subsurface. Simple equipment can be used monitor water table fluctuations or chemical concentrations through time using the existing infrastructure. The time concentration profile may provide indications of the location of DNAPL (Rossabi, 1999).

DIRECT PUSH

Direct push methods are invasive, however they are less disruptive than conventional drilling. Some advantages of direct push methods such as the cone penetrometer are rapid inexpensive access to the subsurface, minimal investigation derived waste, the ability to mount multiple sensors on the access platform and the ability to collect very high-resolution data with depth. These methods are becoming the method of choice at sites where the use of direct push is possible. The principal disadvantages are they can only be used in unconsolidated materials and they are ultimately limited in depth. The cone penetrometer truck has a standard suite of sensors (tip stress, sleeve friction, pore pressure, and electrical conductivity) which can provide a real time indication of soil type with depth at centimeter scale resolution. At SRS we've used these basic tools to produce a map of the topography of an important confining unit that constrains DNAPL movement. We have also combined the soil type sensors with tools to collect depth discrete gas, water, and soil samples. These more targeted samples help focus DNAPL investigations.

Some of the most promising tools for DNAPL characterization have been incorporated with the cone penetrometer test (CPT). Optical probing through a hardened window in the CPT push rod has been used to detect DNAPL either directly by Raman spectroscopy (Rossabi et al., 2000) or inferentially using induced fluorescence measurements of co-constituents dissolved in the DNAPL (Kram, 1998; Rossabi and Nave, 1998). Visual detection of DNAPL has also been accomplished with a high resolution video microscope mounted in a CPT probe (Lieberman and Knowles, 1998). Another important tool developed for a Geoprobe™ system and modified for cone penetrometer use is the Membrane Interface Probe™ (MIP). Using a selectively permeable membrane permitting entry of only volatile organic compounds (VOCs), the tool provides a continuous profile of VOCs through depth (Christy, 1998) Finally, the Ribbon NAPL Sampler (RNS) by FLUTe can be used in conjunction with direct push or drilling methods. It consists of a hydrophobic sorbent liner with an impregnated indicator dye that is deployed in a borehole directly contacting the formation (Riha et al., 2000). DNAPL is wicked into the liner and causes a color change in the dye. The liner is retrieved from the borehole and colored spots and their depths are logged as DNAPL detect locations. This has been the most consistently robust DNAPL characterization technique that we have used to date.

PARTITIONING TRACERS

Chemicals that strongly partition into an organic phase may be used with conservative tracers to determine the presence of NAPLs and, in some cases, the NAPL saturation (Jin et al., 1995). During this test, partitioning and conservative tracers are injected into the formation though a well and recovered in a different well(s). By analyzing the center of mass arrival times of the conservative and reactive tracers, the amount of residual NAPL may be estimated because the retardation of the reactive tracer is attributed to partitioning to the NAPL. This method has been successfully used at many NAPL contaminated sites and has been particularly useful in comparing the performance of clean up methods by pre and post testing. The advantages of this method include the ability to probe a large volume, relatively simple formulas for interpretation, and that they can be used both above and below the water table. The most important issue with this technology is that the tracers can only infer NAPL that is in the tracer flowpath.

This is particularly critical in the vadose zone where organic contaminants from old releases tend to reside in the fine grain soils not easily accessed by tracer gases which prefer to flow through more permeable zones. Other issues with this technology include permitting and waste handling of the generally large volume of tracers required and limiting the error in the empirically determined, partitioning coefficients of the tracers into the NAPL. A recently developed complementary technology that has increased the utility of this method is interfacial area partitioning tracer tests (Kim et al., 1997). This helps determine the surface area of NAPL available for the NAPL partitioning tracer.

Another test using a solubilizing fluid injected through a cone penetrometer rod was conducted at a DNAPL site on the Cape Canaveral Air Station (CCAS). In this test, a small volume (less than 4 liters) of potable water is injected and recovered at a specific depth through a cone penetrometer system to determine hydraulic parameters and baseline concentrations. A similar volume of an alcohol solution or other NAPL solubilizing fluid is then injected and recovered at the same location. The concentrations of the target organic contaminant during the two tests are then compared. If NAPL is encountered, the recovered solubilizing fluid will have significantly greater concentrations of the contaminant than the water recovery. If only aqueous contaminant is present both tests will have similar concentrations in the recovered samples. During the CCAS test, higher concentrations of TCE were measured in the recovered fluid after alcohol injection than in the fluid recovered following water injection, suggesting TCE DNAPL in the test zone of influence. In addition, concentrations of cis dichloroethylene (existing only in the aqueous phase as a byproduct of reductive dechlorination of TCE) remained the same following both injections. This test takes advantage of the high resolution geologic data that the cone penetrometer provides to focus the investigation and minimize waste. Unfortunately this also results in a smaller volume probed and the test has similar NAPL flowpath contact issues as the partitioning interwell tracer test (PITT™).

CONCLUSIONS

Each technology has advantages and disadvantages in its application. Geophysical techniques image a volume and are noninvasive but may lack the resolution to directly characterize a DNAPL-contaminated site. These methods are quite useful in mapping subsurface units that may control the movement of DNAPL. Partitioning or solubilizing tracer tests also probe a large subsurface volume and can detect small pockets of separate phase contaminants but the contaminants must be in the flowpath of the tracers. There may also be permitting and waste issues associated with these methods. Direct sampling or sensing, particularly when applied with direct penetration tools, can offer positive identification of DNAPL at very high vertical resolution but low lateral resolution because the methods do not probe beyond the radius of the borehole. They also require additional boreholes, which increases the cost of characterization.

Ideally, many techniques would be used in a characterization effort because of the complementary information they provide but a limited budget often precludes that choice. The unique features of a specific site will dictate and narrow the list of appropriate tools and the cost of the technologies will further constrain the selection. Finally the order of technology application is important, generally progressing from low cost, less invasive techniques to more specific technologies. By selecting the right combination of technologies, both conventional and innovative, and using these in the frame of an evolving conceptual model of the site, the most effective DNAPL characterization can be obtained.

REFERENCE

Cohen, R. M., and J. W. Mercer. 1993. *DNAPL Site Evaluation*, C. K. Smoley-CRC Press, Boca Raton, Florida.

Christy, T. 1998. "A Permeable Membrane Sensor for the Detection of Volatile Compounds in Soil", in *Proceedings of the Symposium on the Application of Geophysics to Engineering and Environmental Problems*, Chicago, IL.

Daily, W., A. Ramirez, D. LaBrecque, and J. Nitao. 1992. "Electrical Resistivity Tomography of Vadose Water Movement." *Water Resour. Res.* 28 (5): 1429-1442.

Eddy-Dilek, C. A., B. D. Riha, D. Jackson, and J. Rossabi. 1999. *DNAPL Source Zone Characterization of Launch Complex 34, Cape Canaveral Air Station, Florida*. WSRC-TR-99-00024, Westinghouse Savannah River Co., Aiken, SC 29808.

Jin, M., M. Delshad, V. Dwarakanath, D. C. McKinney, G. A. Pope, K. Sepehrnoori, C. Tilburg, and R. E. Jackson. 1995. "Partitioning tracer test for detection, estimation and remediation performance assessment of subsurface nonaqueous phase liquids." *Water Resour. Res.* 31(5):1201-1211.

Kim, H., P. S. C. Rao, and M. D. Annable. 1997. "Determination of effective air-water interfacial area in partially saturated porous media using surfactant adsorption." *Water Resour. Res.* 33(12):2705-2711.

Kram, Mark. 1998. "Use of SCAPS petroleum hydrocarbon sensor technology for real-time indirect DNAPL detection.", *Journal of Soil Contamination, 17(1), 73-86*, 1998.

Lieberman, S.H., and D.S. Knowles. 1998. "Cone penetrometer deployable in situ video microscope for characterizing sub-surface soil properties." *Field Analytical Chemistry and Technology* 2(2):127-132.

Nelson, P.H. and J.E. Kibler. 1996. *Geophysical Logs and Groundwater Geochemistry in the A/M Area, Final Report, Savannah River Site, South Carolina*. USGS Open-File Report 96-75.

Riha, B. D., J. Rossabi, C. A. Eddy-Dilek, and D. Jackson. 2000. "DNAPL Characterization Using the Ribbon NAPL Sampler: Methods and Results", in *Proceedings of the Second International Conference on Remediation of Chlorinated and Recalcitrant Compounds*, Monterey, CA, May 22-25, 2000.

Rossabi, J. and S. E. Nave. 1998. *Characterization of DNAPL using Fluorescence Techniques (U)*, WSRC-TR-98-00125, Westinghouse Savannah River Company, Aiken, SC 29808.

Rossabi, J. 1999. *The influence of atmospheric pressure variations on subsurface soilgas and the implications for environmental characterization and remediation*, Ph.D. thesis, Clemson University, University of Michigan Press.

Rossabi, J., B. D. Riha, J. Haas, C. A. Eddy-Dilek, A. Lustig, M. Carrabba, W. K. Hyde, and J. Bello. 2000. "Field Tests of a DNAPL Characterization System Using Cone Penetrometer-based Raman Spectroscopy." *Ground Water Monitoring and Remediation*, in press.

West, O.R., D. W. Bottrell, C. K. Bayne, R. L. Siegrist, and W. H. Holden. 1996. "A Scientific Basis for Revising Regulatory Holding Times for VOC Water Samples." *Am. Envrion. Lab.*, pp. 15-16.

DNAPL DELINEATION WITH SOIL AND GROUNDWATER SAMPLING

Arun Gavaskar, Stephen Rosansky, Steve Naber, Neeraj Gupta, Bruce Sass,
and Joel Sminchak (Battelle, Columbus, Ohio)
Major Paul DeVane (Air Force Research Laboratory, Tyndall AFB, Florida)
Thomas Holdsworth (U.S. Environmental Protection Agency, Cincinnati, Ohio)

ABSTRACT: Chlorinated solvent contamination consisting primarily of trichloroethylene (TCE) is present at Launch Complex 34 (LC34), Cape Canaveral Air Station, Florida. The presence of dense nonaqueous phase liquid (DNAPL) was confirmed following groundwater and soil sampling for chlorinated volatile organic compounds (CVOCs) during preliminary site characterization. Detailed site characterization was conducted in two stages. The Stage 1 soil sampling in the suspected DNAPL zone was used to prepare a Stage 2 soil sampling plan based on a statistical design to map the DNAPL source zone boundaries, determine the horizontal and vertical distribution of the DNAPL, and estimate the total TCE and DNAPL masses. Groundwater (organic and inorganic) analysis and hydraulic parameter measurements were conducted to track the progress of the remediation, evaluate DNAPL migration potential, and evaluate aquifer quality before and after treatment. Depth-discrete soil and groundwater sampling was found to be an efficient way of characterizing the DNAPL source zone to enable the design of the remediation technology application, as well as to track remediation progress and effectiveness.

INTRODUCTION

The Interagency DNAPL Consortium (IDC), a consortium consisting of U.S. Department of Energy (DOE), U.S. Department of Defense of DoD), U.S. Environmental Protection Agency (EPA), and the National Aeronautic and Space Administration (NASA) are conducting a demonstration of three in situ remediation technologies – six-phase heating (SPH™), chemical (permanganate) oxidation, and steam injection. Battelle, Columbus, Ohio and U.S. EPA conducted the detailed site characterization at LC34 and are conducting an independent performance assessment of the applicability of the three technologies.

Preliminary site characterization by Westinghouse Savannah River Co. (WSRC) at LC34 showed the presence of CVOCs, primarily trichloroethylene (TCE) and cis-1,2 dichloroethylene (DCE), associated with the soil and groundwater. WSRC mapped and quantified the CVOC distribution in the whole LC34 area and confirmed the presence of DNAPL on the north side of the Engineering Service Building (ESB) at LC34 (Eddy-Dilek et al., 1998). Subsequently, Battelle conducted detailed characterization of the DNAPL source zone near the ESB to determine the exact boundaries of the DNAPL for remediation, map the horizontal and vertical distribution of DNAPL near the

ESB, and estimate the total TCE and DNAPL mass in three plots designated for the three technologies being demonstrated (Figure 1). The local stratigraphy, hydraulic properties, and aquifer geochemistry near the ESB were also determined to assist in the application of the remediation technologies.

MATERIALS AND METHODS

In Stage 1 of the detailed characterization in and around the north side of the ESB (Battelle, 1999a), groundwater samples were collected from 3 existing well clusters (IW-1, IW-17, and IW-27) and 12 new well clusters (PA-1 to PA-12) that were installed (see Figure 1). A cone penetrometer testing (CPT) rig was used to collect soil samples, install wells, and map the stratigraphy at the 12 new locations. Soil core samples were collected from locations both inside and outside the ESB building (north end) because indications were that the DNAPL zone extended partly under the building. In addition to mapping the local CVOC concentrations, the Stage 1 detailed characterization data were used to determine the horizontal and vertical variability of the CVOC distribution. These variability data were used as the basis for a statistical design of soil sampling locations in the three treatment plots. These sampling locations were useful for:

- Mapping the total TCE and DNAPL distribution in the three plots so that treatment could be focused accordingly
- Comparing before- and after-treatment TCE concentrations at paired locations inside each plot to determine the total TCE mass removed by the treatment

An unaligned systematic sampling scheme was used to determine sampling locations and number of samples for Stage 2 soil characterization (Battelle, 1999b). This sampling scheme is shown in Figure 2 and was used because it provides the best horizontal and vertical coverage of the plot, given the variability experienced in Stage 1. A 4 x 3 grid was demarcated in each plot and 12 samples were collected from the locations shown in Figure 2. Initially, additional soil cores were planned to be collected from extended rows of grid cells surrounding each plot, the objective being to evaluate the potential for DNAPL migration from the plots due to the treatments. However, due to resource limitations, soil coring in Stage 2 was limited to the inside of the plots. Monitoring well clusters installed around the plots served as monitors for evaluating DNAPL migration.

Because of the high variability in TCE concentrations with depth, soil core analysis involved collection, extraction, and analysis of every 2-foot depth interval in the entire soil column from ground surface to aquitard. Each 2-foot soil core sample was split vertically into three sections, and one section was extracted and analyzed completely. In this manner, the TCE distribution across the entire vertical soil column at each of the 12 locations in a plot was mapped. This is different from many previous studies, in which only a part of the soil column is sampled and/or small aliquots of soil are isolated from each sample for extraction. A modification of EPA Method 5035 was used to extract each soil sample. Because of the large number of samples (approximately 800 soil samples) involved, the need to extract larger aliquots of soil, and the to enable the conduct of the extractions in the field, a less hazardous solvent, methanol, was used for extraction. Matrix and surrogate spike information indicates that approximately 80% extraction efficiency was obtained by this field method. The extracts were sent to a certified off-site laboratory, which analyzed them by EPA Method 8260. Based on the porosity and adsorptive properties of the aquifer, as well as the solubility of TCE in the pore water, approximately 300 mg/kg of TCE was designated as the threshold concentration between dissolved-phase and DNAPL for this site.

In Stage 2, 8 more well clusters were installed in the three plots for more localized evaluation. Groundwater sampling was conducted with teflon tubing and peristaltic pump. A flowthrough cell was used to measure field parameters, such as dissolved oxygen (DO), oxidation-reduction potential (ORP), and pH. Micropurging, which involves pumping of small volumes of water at low flow rates (Kearl et al., 1994), was done in each well until field parameters stabilized. Early rinsate blanks indicated that despite several repetitions, the standard decontamination procedure (detergent and water rinse) was not sufficient to decontaminate Teflon tubing that had contacted groundwater containing TCE levels close to or above saturation. A methanol rinse was added but failed to adequately clean the tubing. After testing several different decontamination procedures in the laboratory and field, it was decided that fresh tubing would be used for every single groundwater sample during the demonstration. Groundwater samples collected were analyzed for CVOCs, cations, and anions. The objective was to compare before- and after-treatment water quality in the plots.

Hydraulic properties measured in the three plots and their vicinity included porosity, gradient (water levels), and conductivity (slug tests). The objective was to track changes in porosity and conductivity (as may happen, for example if manganese dioxide builds up in the soil pores in the oxidation plot), as well as evaluate any strong gradients generated during remediation that could potentially cause DNAPL migration from the plots.

RESULTS AND DISCUSSION

Figure 3 shows the stratigraphy that was mapped in the three treatment plots to assist in the remediation. At the LC34 site, the water table occurs close to the ground surface and has varied between 0 to 5 ft below ground surface (bgs) during the demonstration. The confining layer, called the lower clay unit (LCU), is approximately 45 ft bgs. In the predominantly sandy aquifer, an intermediate silty layer, called the middle fine-grained unit (MFGU), occurs at approximately 25 to 30 ft bgs. The portions of the aquifer above and below the MFGU are called the upper sand unit (USU) and lower sand unit (LSU). Localized stratigraphic mapping is important because much of the DNAPL was found at depths close to the MFGU and LCU, indicating that these fine-grained act to retard and retain the downward-migrating DNAPL (see Figure 4 for vertical DNAPL distribution). As seen in Figure 3, the thickness of the MFGU varies considerable across the targeted treatment zone and technology application strength (heating strength or oxidant dosage) had to focus on these regions.

The horizontal distribution of the TCE and DNAPL is shown in Figure 5. As mentioned above, 300 mg/kg of TCE in soil was used as the cutoff between dissolved-phase and DNAPL. The estimated total TCE and DNAPL masses in the three plots are shown in Table 1. These mass estimates were used by the three technology vendors to design and cost their applications.

TABLE 1. Estimated total TCE and DNAPL masses near the ESB at LC34

SPH™ *Plot*		*Steam Injection Plot*		*Oxidation Plot*	
Total* TCE (kg)	Estimated DNAPL (kg)	Total* TCE (kg)	Estimated DNAPL (kg)	Total* TCE (kg)	Estimated DNAPL (kg)
11,313	10,490	13,077	11,713	6,122	5,039

* Includes dissolved-phase and solvent-phase TCE

Figure 6 shows the distribution of many of the groundwater parameters measured at the site. In its native state, the aquifer is mostly anaerobic and flow is almost stagnant (relatively flat gradient). Higher chloride concentrations are present deeper in the aquifer due to the proximity to the sea. The aquifer is relatively high in alkalinity and dissolved solids and the pH is close to neutral. This information was used to design the remediation application.

A combination of depth-discrete groundwater and soil sampling was found to be a useful tool for identifying the DNAPL source zone, mapping the horizontal and vertical extent of the DNAPL, estimating the DNAPL mass requiring treatment, and assessing the effectiveness of the remediation. Hydraulic gradient measurements and groundwater sampling provided real time information on remediation progress during the treatment, and are expected to enable the determination of any changes in aquifer properties and/or composition due to treatment.

At the time of this paper submittal, two of the technologies (SPH™ and chemical oxidation) were operating in the two outer plots. Groundwater sampling and hydraulic measurements during the demonstration were used to track remediation progress and verify DNAPL migration potential from the plots. The plan is to collect groundwater and soil samples after the treatments have been concluded. Soil samples will again be collected from the same grid cells in each plot to compare the reduction in TCE concentrations at paired locations. The steam injection demonstration will subsequently begin in the center plot.

ACKNOWLEDGEMENTS

Battelle would like to acknowledge the other IDC members Skip Chamberlain (DOE) and Jackie Quinn (NASA) for their support during the site characterization and performance assessment at LC34. Stan Lynn and John Thompson from Tetra-Tech EMI provided important field support. DHL

Analytical and Quanterra Environmental Services provided analytical support during the demonstration.

REFERENCES

Eddy-Dilek, C., B. Riha, D, Jackson, J. Rossabi, and J. Consort. 1998. *DNAPL Source Zone Characterization of Launch Complex 34, Cape Canaveral, Florida.* Prepared by Westinghouse Savannah River Co. and MSE Technology Applications, Inc. for the IDC, December 23, 1998.

Battelle, 1999a. Performance Assessment Site Characterization for the Interagency DNAPL consortium, Launch Complex 34, Cape Canaveral Air Station, Florida. Prepared by Battelle for Air Force Research Laboratory, August, 1999.

Battelle, 1999b. *Pre-Demonstration Assessment of the Treatment Plots at Launch Complex 34, Cape Canaveral, Florida, Part 1: Soil Analysis and Field Measurements.* September 1999.

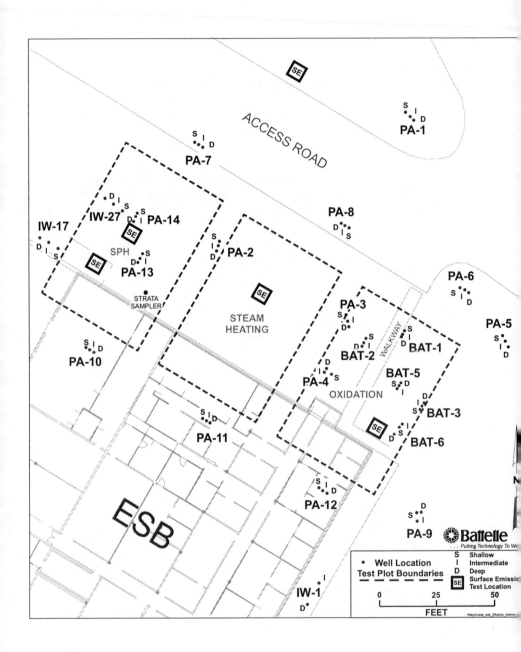

Figure 1. Three Plots Designated for Demonstration of Three DNAPL Remediation Technologies at LC34

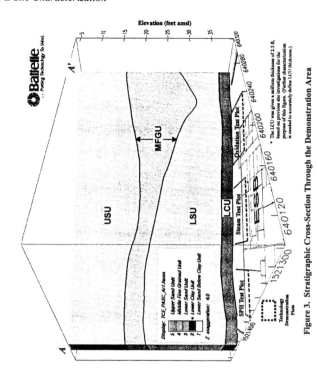

Figure 3. Stratigraphic Cross-Section Through the Demonstration Area

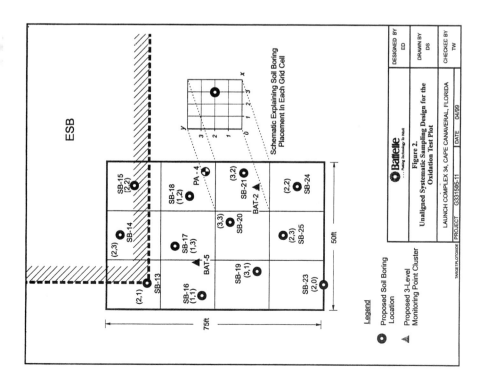

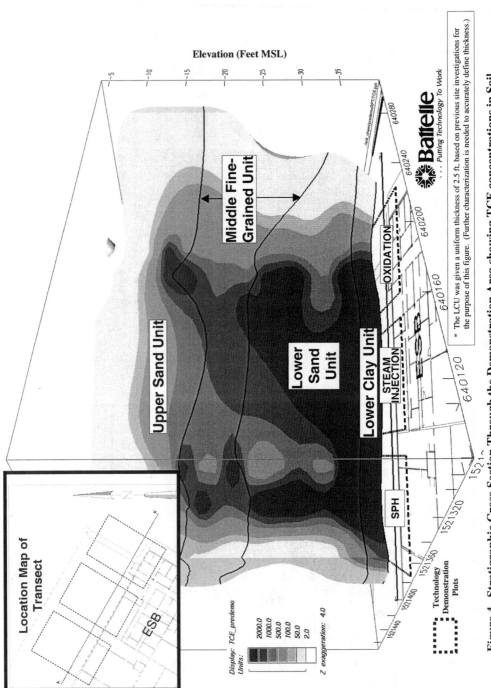

Figure 4. Stratigraphic Cross-Section Through the Demonstration Area showing TCE concentrations in Soil

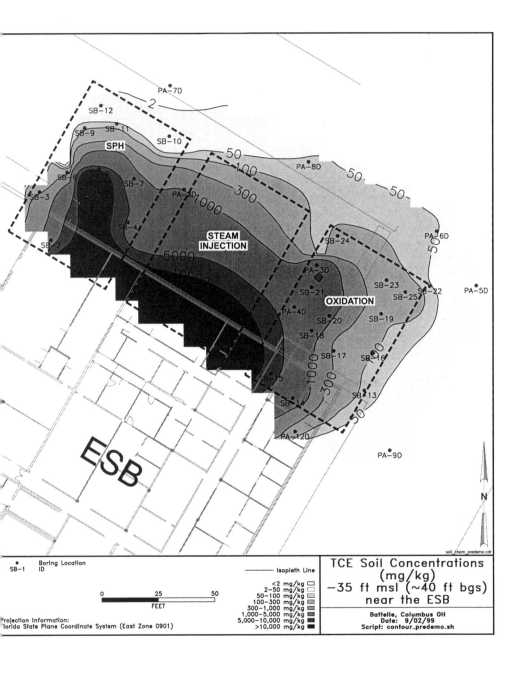

Figure 5. TCE Concentrations (mg/kg) in Soil at ~40 ft bgs near the ESB

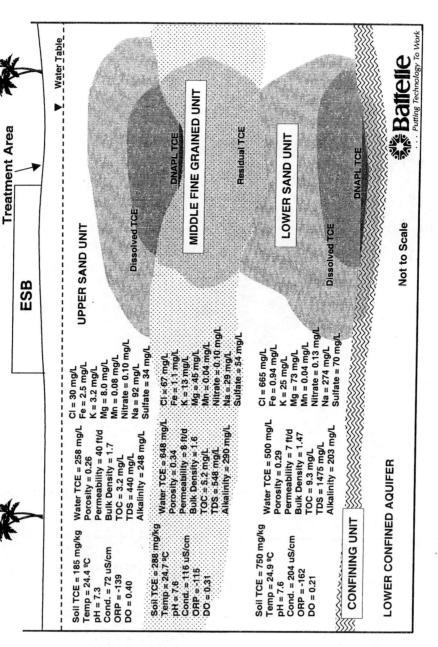

Figure 6. Groundwater Constituents in the Aquifer Near the ESB at LC34

COMPARISON OF DNAPL SITE CHARACTERIZATION APPROACHES

Mark L. Kram (Bren School of Environmental Science and Management, UCSB)
Arturo A. Keller (Bren School of Environmental Science and Management, UCSB)
Joseph Rossabi (Westinghouse Savannah River Company)
Lorne G. Everett (Institute for Crustal Studies, UCSB)

Abstract: The current lack of appropriate methods for detecting and delineating widely dispersed micro-globules of dense non-aqueous phase liquids (DNAPLs) has been identified as one of the most significant challenges limiting effective cleanup of sites contaminated with these pollutants. This study compares many of the approaches and methods currently used to detect and delineate DNAPL contaminant source zones. In addition, a cost comparison is generated using a contrived conceptual site exhibiting particular sets of physical characteristics. The objective is to determine which options are best to pursue based on known site characteristics, method performance capabilities, and method costs. Selected candidate DNAPL characterization methods are grouped into sets of approaches that represent site management options for achieving cost-effective DNAPL source zone characterization. Inherent in these approaches is the goal of identifying and quantifying physical and chemical site characteristics required for effective remediation. We compare the different approaches based on the level of chemical and hydrogeologic resolution, associated costs, and the need for additional data requirements. Our findings can be used to assist with selection of appropriate site remediation management options.

INTRODUCTION

Contamination of soils and groundwater by the release of dense non-aqueous phase liquids (DNAPLs), including halogenated solvents, has posed serious environmental problems for many years. In order to be able to remediate a site contaminated with DNAPLs, it is necessary to remove or isolate undissolved (non-aqueous) product remaining in the subsurface. Failure to remove residual (held under capillary forces) or free phase (mobile) product may result in continued, long-term contamination of the surrounding groundwater. The marginally soluble organic contaminants can partition into the aqueous phase at rates slow enough to continue to exist as a non-aqueous phase, yet rapid enough to render water supplies a threat to public health. DNAPLs can migrate to depths well below the water table. As they migrate, they can leave behind trails of micro-globules in the pore spaces of the soil matrix, which effectively serve as long-term sources of groundwater pollution. Current conceptual DNAPL transport models suggest that when sinking free-phase DNAPL encounters a confining layer (e.g., a competent clay or bedrock zone), it can accumulate, or "pool", and spread laterally until it encounters a fracture or an alternative path of relatively low flow resistance towards deeper zones. In addition, globules can enter pores and be held as a residual phase in capillary suspension. This complex mode of subsurface transport results in unpredictable heterogeneous distribution of non-aqueous product that is difficult to delineate. The current lack of appropriate methods for detecting and delineating widely dispersed micro-globules of DNAPL product has been identified as one of the most significant challenges limiting effective cleanup of sites contaminated with these pollutants (Feenstra *et al.*, 1996).

This paper will describe and compare many of the best approaches and methods currently used to detect and delineate DNAPL contaminant source zones. In addition, a cost comparison will be generated using a contrived conceptual site exhibiting a particular set of physical characteristics. The objective is to determine which site characterization approaches are best to pursue based on known site characteristics, method performance capabilities, and method costs. A distinction between specific methods and site management approaches will be used in the cost analysis. An approach signified by a method descriptor (such as "soil gas survey" or "surface geophysics", etc.) implies that the approach includes the method as part of the overall characterization effort. Selected candidate methods will be grouped into sets of approaches that represent site management options for achieving cost-effective DNAPL source zone characterization. Inherent in these characterization approaches will be the goal of identifying and quantifying physical and chemical site characteristics that will lead towards effective remediation alternatives. Approach comparisons based on the level of chemical and hydrogeologic resolution, associated costs, and the need for additional data requirements will be generated to assist with selection of appropriate site remediation management options.

Environmental characterization efforts for contaminated sites typically evolve through a series of stages. Initially no information is available. We will refer to this stage as t_0. At t_1, some preliminary (generally non-intrusive) information becomes available which indicates the potential for risks associated with contaminant exposure. This information would include items typically contained in a Preliminary Site Assessment. At t_2, data collection activities related to subsurface characterization are sufficient to initiate design of a remediation system. At t_3, the site is considered remediated and monitoring is established to determine whether there is further risk. At t_4, monitoring ceases and regulatory closure is achieved, thereby requiring no further action. The approaches discussed in this paper are comprised of multiple methods applied in a logical sequence with the goal of reaching stage t_2.

METHODS

Comparable cost and performance data for DNAPL site characterization methods and approaches is limited. Rarely are several methods compared to each other on a systematic basis at the same site. Typically, when data is available for a particular approach or method, it is usually compared to a set of confirmation data collected and analyzed using standardized field laboratory methods. The data collection locations for confirmation samples are typically dictated by previous results. For instance, when one uses a field screening technique, confirmation samples are collected from locations identified as polluted or clean based on the field screening method results. Since each method and approach varies in terms of spatial resolution and completeness with respect to requirements for remedial design, corresponding confirmation approaches will also vary. Due to the lack of comparable cost data, the lack of resources for conducting method comparisons in the field under various scenarios, and the differences associated with confirmation approaches anticipated for particular methods, the authors evaluated various DNAPL site characterization methods and approaches using a contrived model site.

A "unit model scenario" (UMS) was selected to describe site characteristics for the cost comparison. The UMS was comprised of unconsolidated sediment in a cylinder of 15.2 m (50-ft) radius and 30.4 m (100-ft) depth with a water table at 4.6 m (15-ft) below ground surface and a 1000-l (264-gal) DNAPL product release volume. The scenario was representative of the shallow groundwater case where DNAPL has already advanced below the water table. For cost comparisons, it was assumed that all releases were initiated at the same time and within 10 years of the initial investigation. Dimensions presented in the model scenario were representative, but by no means all-inclusive. Cost and field time estimates can be adjusted by normalizing (e.g., based on depth, area, or estimated contaminant volume) or by adjusting and fine-tuning the assumptions presented for each approach (see Kram *et al.*, 2000, for more information). Speculation about actual configurations of NAPL source zones is beyond the scope of this effort.

DESCRIPTIONS OF DNAPL SITE CHARACTERIZATION TECHNIQUES

The techniques described in this section were selected because they have been used at several sites to identify DNAPL source zones. In addition, each of the methods described has demonstrated potential for successful DNAPL plume delineation, either directly or indirectly. Some of the methods have been extensively tested (e.g., sample collection and analysis, soil gas surveys, seismic surveys, and other geophysical surveys), while others are considered relatively new techniques (e.g., FLUTe, ultraviolet (UV) fluorescence using a cone penetrometer, and precision injection extraction (PIX)). Table 1 lists each of these characterization methods and identifies pertinent references for obtaining additional information. Kram *et al.* (2000, in print) describes each of the methods in greater detail.

TABLE 1. DNAPL site characterization methods.

Methods	References
1) Baseline Methods: Field Observations	Cohen and Mercer, 1993; Pankow and Cherry, 1996
Baseline Methods: Chemical Analysis of Soil, Rock and Water from Coincidental Samples (Including Fault Planes)	Cohen *et al*, 1992; Cohen and Mercer, 1993; Pankow and Cherry, 1996; MSE, 2000
Baseline Methods: Visual Evidence	Cohen and Mercer, 1993; Pankow and Cherry, 1996
Baseline Methods: Enhanced Visual Identification: Shake-Tests	Cohen *et al.*, 1992; Cohen and Mercer, 1993; Pankow and Cherry, 1996
Baseline Methods: Enhanced Visual I.D.: UV Fluorescence w/Portable Light	Cohen *et al.*, 1992; Pankow and Cherry, 1996
Baseline Methods: Enhanced Visual I.D.: Dye Addition w/Sudan IV or OIL Red O	Cohen *et al.*, 1992; Cohen and Mercer, 1993; Pankow and Cherry, 1996
Baseline Methods: Vapor Analysis while Sampling Sediments or Drilling	Cohen and Mercer, 1993; Pankow and Cherry, 1996
Baseline Methods: Drilling Water Analysis	Taylor and Serafini, 1988; Cohen and Mercer, 1993; Pankow and Cherry, 1996
Baseline Methods: Observation Wells	Cohen and Mercer, 1993; Pankow and Cherry, 1996
Baseline Methods: Test Pits	Pankow and Cherry, 1996
2) Soil Gas Surveys	Marrin, 1988; Marrin and Kerfoot, 1988; Cohen and Mercer, 1993
3) Partitioning Interwell Tracer Tests	Jin *et al.*, 1995; Nelson and Brusseau, 1996; Burt *et al.*, 1998; Payne *et al.*, 1998; Meinardus *et al.*, 1998; Knox *et al.*, 1998; Annable *et al.*, 1998; Nelson *et al.*, 1999; Dwarakanath *et al.*, 1999; Wise, 1999; Yoon *et al.*, 1999.
4) Radon Flux Rates	Semprini *et al.*, 1998
5) Back-Tracking Using Dissolved Concentrations in Wells	Feenstra and Cherry, 1988; Feenstra *et al.*, 1991; Newell and Ross, 1991; Cohen and Mercer, 1993; Johnson and Pankow, 1992; Anderson *et al.*, 1992; Pankow and Cherry, 1996
6) Surface Geophysics	Cohen and Mercer, 1993; Pankow and Cherry, 1996; Adams *et al.*, 1998; Sinclair and Kram, 1998
7) Subsurface Geophysics	Brewster *et al.*, 1992; Cohen and Mercer, 1993; Pankow and Cherry, 1996
8) CPT Methods: Permeable Membrane Sensor; Membrane Interface Probe (MIP)	Christy, 1998
CPT Methods: HydroSparge	Davis *et al.*, 1997; Davis *et al.*, 1998
CPT Methods: Florescence (e.g., Laser Induced Fluorescence, LIF) Techniques	Kram, 1996; Kram, 1997; Kram *et al.*, 1997; Kram, 1998; Keller and Kram, 1998; Kram *et al.*, 2000; MSE, 2000; Lieberman *et al.*, 2000
CPT Methods: GeoVis	Lieberman and Knowles, 1998; Lieberman *et al.*, 2000
CPT Methods: LIF/ GeoVis	Lieberman and Knowles, 1998; Lieberman *et al.*, 1998; Lieberman *et al.*, 2000
CPT Methods: Raman Spectroscopy	Mosier-Boss *et al.*, 1997; Rossabi *et al.*, 2000 (in print)
CPT Methods: LIF/Raman	Kenny *et al.* (1999)
CPT Methods: Electro-Chemical Sensor Probe	Adams *et al.*, 1997
CPT Methods: Waterloo (Ingleton) Profiler	Pitkin, 1998; Sudicky, 1986
CPT Methods: Cosolvent Injection/Extraction; Precision Injection/Extraction (PIX) Probe	Looney *et al.*, 1998; MSE, 2000
9) Flexible Liner Underground Technologies Everting (FLUTe) Membrane	MSE, 2000

COST ANALYSIS

This section presents cost analyses for each of the methods and approaches presented above and described in Kram *et al.*, 2000. To generate a useful cost comparison, several assumptions were required. These assumptions are presented in Kram *et al.*, 2000. Each approach was compared to a common baseline approach that consists of sample collection from the surface and from consecutive 1.5-m (5-ft) depth intervals. We do not mean to imply that a 1.5-m (5-ft) level of resolution is valid for all sites; rather we consider this a typical sampling increment. Although commonly used, the chance for misidentifying DNAPL ganglia and microglobules using this type of approach is very high. In addition, penetration of zones containing free-phase DNAPL could lead to vertical migration of contaminants to deeper zones, exacerbating the problems associated with the release.

It is important to recognize that each method presents specific advantages and disadvantages and that, due to the nature of each method and the sequence with which it can be applied in the overall site management process, direct comparisons involve some uncertainty. A project manager that knows little about the location of DNAPL at a site yet is interested in the most cost-effective approach, must place each candidate method in the proper context within the characterization process. Comparison of characterization components in isolation tends to bias the cost estimate, thereby rendering the comparison fallible. In an attempt to normalize the comparison, each method is evaluated in a manner consistent with the niche fulfilled. It is assumed that little is known and that the project manager wants to obtain enough information to determine whether the site is clean or how to properly design a remediation system based on specific site constraints. As mentioned above, a distinction between specific methods and site management approaches is employed. Therefore, wherever possible, the approaches described below include not only the specific methods of interest, but also confirmation methods, and in some cases preliminary characterization efforts utilizing additional methods. The authors opted for this process since we believe that comparison of components alone would not be sufficient to compare the management options.

A detailed presentation of the cost estimates and savings estimates for each approach can be found in Kram *et al.*, 2000. Figure 1 displays the cost comparisons graphically.

DISCUSSION AND CONCLUSIONS

The least expensive approaches include several CPT sensor approaches, such as Fluorescence and MIP, and the FLUTe approach. In this example, the FLUTe approach was installed with a CPT device. The Fluorescence and MIP approaches must always include confirmation efforts, either by use of conventional analyses or by coupling to additional sensors such as the GeoVis. However, MSE (2000) believe that the FLUTe approach may not require chemical confirmation once a larger database has been generated. If supported by regulators, this will substantially reduce the costs (by close to $6000) associated with the FLUTe approach. The FLUTe approach may be more definitive with respect to identifying DNAPL source zones. While the Fluorescence and MIP approaches generate soil classification data, the FLUTe approach will either require that lithology sensors are operating during the preliminary pushes, or that additional laboratory tests be conducted on soil samples to determine soil type and properties. Several additional approaches, including soil gas, Hydrosparge, GeoVis, Fluorescence-GeoVis, Raman, Fluorescence-Raman, and the Waterloo Profiler are very competitive (ranging from $20,000 to $40,000).

The most expensive approach for this scenario is the PITT survey. While this approach yields detailed hydrologic information and NAPL volume estimates on a localized scale, water treatment costs associated with hydraulic control, and costs associated with preliminary site characterization and setup (e.g., aquifer testing, well installation, etc.) can be very high. If a site has been well characterized and wells are properly installed and screened in optimal locations, the PITT approach can be a useful endeavor. PITT approaches for evaluation of remediation effectiveness have been successfully demonstrated with remarkable mass removal estimates (Meinardus *et al.*, 1998). During one particular test conducted at Hill Air Force Base, Utah, PITT estimated that approximately 1,310 liters (346 gallons) of residual DNAPL remained in a test area prior to removal with use of a surfactant. A post-remediation PITT indicated that 1,291 liters (341 gallons) had been recovered, with approximately 19 liters (5 gallons) remaining in the swept volume. The effluent treatment system recorded 1,374 liters (363 gallons) recovered.

As briefly outlined in Table 1 of Kram *et al.* (2000), each method has specific advantages and disadvantages. Several methods can be complementary in an overall site management plan, each serving a particular niche. This can be considered a "hybrid" approach, whereby the strengths of individual characterization components are exploited at the most appropriate and logical times in the site management process. For example, under conditions used in this exercise, one can initially screen the site with a FLUTe or LIF/GeoVis method, then analyze confirmation samples. After determining the location of the DNAPL source zone, discreetly screened or multi-level wells can be installed and a Radon flux rate or PITT survey can be used to estimate the amount of NAPL present. The number of available method combinations and potential options are extensive. Cost estimates presented in Kram *et al.* (2000) and summarized in this paper can be used to estimate anticipated costs for these hybrid management strategies.

This paper compares many of the methods and approaches currently used to detect and delineate DNAPL contaminant source zones. General performance comparisons (presented in detail in Kram *et al.*, 2000) were generated to identify potential site management considerations required to reach a level of site understanding adequate to initiate remediation design efforts. In particular, advantages and disadvantages for several methods were presented. In addition, characterization approach cost comparisons for a contrived conceptual site exhibiting particular sets of physical characteristics were generated. Perhaps the most important issue raised deals with the recognition that each candidate method must be placed in its proper context within the characterization process. Approach comparisons based on the level of chemical and hydrogeologic resolution, associated costs, and the need for additional data requirements have been generated to assist with selection of appropriate site remediation management options.

ACKNOWLEDGEMENTS
We are grateful for all the assistance received from the many dedicated researchers and associates in the environmental industry.

CAVEAT
The assumptions stated in this paper are the opinions of the authors' and do not constitute endorsements of particular approaches or methods, nor are they representative of the opinions of their organizations. The cost data used in this analysis was synthesized from a variety of sources including the authors' experiences, several commercial vendors, consultants and government employees within the environmental industry.

REFERENCES
Adams, Mary-Linda, Brian Herridge, Nate Sinclair, Tad Fox, Chris Perry, 1998. "3-D Seismic Reflection Surveys for Direct Detection of DNAPL", in *Proceedings for the First International Conference on Remediation of Chlorinated and Recalcitrant Compounds, Non-Aqueous-Phase Liquids*, Monterey, CA, May 18-21, 1998, pp. 155-160.

Adams, Jane W., William M. Davis, Ernesto R. Cespedes, William J. Buttner, and Melvin W. Findlay, 1997. "Development of Cone Penetrometer Electro-Chemical Sensor Probes for Chlorinated Solvents and Explosives", in Proceedings for a Specialty Conference on Field Analytical Methods for Hazardous Wastes and Toxic Chemicals, January 29-31, 1997, Las Vegas Nevada, pp. 667-670.

Anderson, Michael R., Richard L. Johnson, and James M. Pankow, 1992. "Dissolution of Dense Chlorinated Solvents into Groundwater. 3. Modeling Contaminant Plumes from Fingers and Pools of Solvent", *Environmental Science and Technology*, v. 26, pp. 901-908.

Annable, M.D., J.W. Jawitz, P.S.C. Rao, D.P. Dai, H. Kim, and A.L. Wood, 1998. "Field Evaluation of Interfacial and Partitioning Tracers for Characterization of Effective NAPL-Water Contact Areas", *Ground Water*, vol. 36, no. 3, pp. 495-502, June 1998.

Burt, Ronald A., Robert D. Norris, and David J. Wilson, 1998. "Modeling Mass Transport Effects in Partitioning Inter-Well Tracer Tests", in *Proceedings for the First International Conference on Remediation of Chlorinated and Recalcitrant Compounds, Non-Aqueous-Phase Liquids*, Monterey, CA, May 18-21, 1998, pp. 119-124.

Brewster, M.L., A.P. Annan, J.P. Greenhouse, G.W. Schneider, and J.D. Redman, 1992. "Geophysical Detection of DNAPLs: Field Experiments", in *Proceedings: International Association of Hydrogeologists Conference*, Hamilton, Ontario, May 1992, pp. 176-194.

Christy, Thomas M., 1998. "A Permeable Membrane Sensor for the Detection of Volatile Compounds in Soil", *Proceedings for the Annual Meeting of the Environmental and Engineering Geophysical Society, March 22-26, 1998*, Chicago, Illinois, p.65-72.

Cohen, Robert M. and James W. Mercer, 1993. *DNAPL Site Evaluation*, Published by C.K. Smoley.

Cohen, R.M, A.P. Bryda, S.T. Shaw, and C.P. Spalding, 1992. "Evaluation of Visual Methods to Detect NAPL in Soil and Water", *Ground Water Monitoring Review*, v. 12, no. 4, pp. 132-141.

Davis W.M., J.F. Powell, K. Konecny, J. Furey, C.V. Thompson, M. Wise, and G. Robitaille, 1997. "Rapid In-Situ Determination of Volatile Organic Contaminants in Groundwater Using the Site Characterization and Analysis Penetrometer System", in *Proceedings for Field Analytical Methods for Hazardous Wastes and Toxic Chemicals Conference*, VIP 71, Air and Waste Management Association, Pittsburgh, PA, p. 464-469.

Davis, W. M., M.B. Wise, J.S. Furey, and C.V. Thompson, 1998. "Rapid Detection of Volatile Organic Compounds in Groundwater by In Situ Purge and Direct-Sampling Ion-Trap Mass Spectrometry", *Field Analytical Chemistry and Technology*, v. 2, no. 2, pp. 89-96.

Dwarakanath, V., N. Deeds, and G.A. Pope, 1999. "Analysis of Partitioning Interwell Tracer Tests", *Environmental Science and Technology*, v.33, pp.3829-3836.

Feenstra, S., D.M. Mackay, and J.A. Cherry, 1991. "Presence of Residual NAPL Based on Organic Chemical Concentrations in Soil Samples", *Ground Water Monitoring Review*, v. 11, no. 2, pp. 128-136.

Feenstra, S., J.A. Cherry, and B.L. Parker. 1996. "Conceptual Models of the Behavior of Dense Non-Aqueous Phase Liquids (DNAPLS) in the Subsurface", *In: Dense Chlorinated Solvents and other DNAPLs in Groundwater: History, Behavior, and Remediation*, Waterloo Press, 53-88.

Feenstra, S. and J.A. Cherry, 1988. "Subsurface Contamination by Dense Non-Aqueous Phase Liquid (DNAPL) Chemicals", in *Proceedings: International Groundwater Symposium, International Association of Hydrogeologists*, May 1-4, Halifax, Nova Scotia, pp. 62-69.

Jin, M., M. Delshad, V. Dwarakanath, D.C. McKinney, G.A. Pope, K. Sepehrnoori, C.E. Tilburg, and R.E. Jackson, 1995. "Partitioning Tracer Test for Detection, Estimation and Remediation Performance Assessment of Subsurface Nonaqueous Phase Liquids", Water Resources Research, v. 31(5), pp. 1201-1211.

Johnson, Richard L., and James M. Pankow, 1992. "Dissolution of Dense Chlorinated Solvents into Groundwater. 2. Source Functions for Pools of Solvents", Environmental Science and Technology, v. 26, pp. 896-901.

Keller, Arturo A. and Mark L. Kram, 1998. "Use of Fluorophore/DNAPL Mixtures to Detect DNAPLs In-Situ", in *Nonaqueous-Phase Liquids, Remediation of Chlorinated and Recalcitrant Compounds*, Battelle Press, eds. Godage B. Wickramanayake and Robert E. Hinchee, pp.131-136.

Kenny, Jonathan E., Jane W. Pepper, Andrew O. Wright, Yu-Min Chen, Steven L. Schwartz, and Charles G. Skelton, 1999. *Subsurface Contamination Monitoring Using Laser Fluorescence*, edited by Katherine Balshaw-Biddle, Carroll L. Oubre, and C. Herb Ward, Lewis Publishers, 160 pages.

Knox, Robert C, David A. Sabatini, Micah Goodspeed, Mark Hasegawa, and Leon Chen, 1998. "Hydraulic Considerations for Advanced Subsurface Characterization and Remediation Technologies", *IAHS Publication*, no. 250, pp. 391-399.

Kram, Mark L., 1996. "Framework for Successful SCAPS Deployment", in *Proceedings of the Sixth Annual AEHS West Coast Conference on Contaminated Soils and Groundwater: Analysis, Fate, Environmental and Public Health Effects, and Remediation*, Newport Beach, CA, 1996.

Kram, Mark L., 1997. "Use of SCAPS Petroleum Hydrocarbon Sensor Technology for Real-Time Indirect DNAPL Detection", in *_Proceedings of the Seventh Annual AEHS West Coast Conference on Contaminated Soils and Groundwater: Analysis, Fate, Environmental and Public Health Effects, and Remediation*, Oxnard, CA, 1997.

Kram, Mark L., Marlene Dean, and Rod Soule, 1997. "The ABCs of SCAPS", *Soil and Groundwater Cleanup*, May 1997, pp. 20-22.

Kram, Mark L., 1998. "Use of SCAPS Petroleum Hydrocarbon Sensor Technology for Real-Time Indirect DNAPL Detection", *Journal of Soil Contamination*, 1998, Volume 7, No. 1, pp. 73-86.

Kram, Mark L., Stephen H. Lieberman, Jerome Fee, and Arturo A. Keller, 2000 (in print). "Use of LIF for Real-Time In-Situ Mixed NAPL Source Zone Detection at a Plating Shop Waste Disposal Site", *Ground Water Monitoring and Remediation*, 2000, in print.

Kram, Mark L., Arturo A. Keller, Joseph Rossabi, and Lorne G. Everett, 2000 (in print). "A Cost and Performance Comparison of Selected DNAPL Site Characterization Methods and Approaches", *Ground Water Monitoring and Remediation*, 2000, in print.

Lieberman, S.H., G.A. Theriault, S.S. Cooper, P.G. Malone, R.S. Olsen and P.W. Lurk, 1991. "Rapid, Subsurface, In-Situ Field Screening of Petroleum Hydrocarbon Contamination Using Laser Induced Fluorescence over Optical Fibers", in *Proceedings: Second International*

Symposium on Field Screening methods for Hazardous Wastes and Toxic Chemicals, U.S. Environmental Protection Agency, Las Vegas, NV, p. 57-63.

Lieberman, Stephen H. and David S. Knowles, 1998. "Cone Penetrometer Deployed In Situ Video Microscope for Characterizing Sub-Surface Soil Properties", *Field Analytical Chemistry and Technology*, v. 2, no. 2, pp. 127-132.

Lieberman, Stephen H., Greg W. Anderson and Andrew Taer, 1998. "Use of a Cone Penetrometer Deployed Video-Imaging Systems for In Situ Detection of NAPLs in Subsurface Soil Environments", in Proceedings: *1998 Petroleum Hydrocarbon and Organic Chemicals in Ground Water: Prevention, Detection, and Remediation, Conference and Exposition*, National Ground Water Association, Houston, Westerville, OH, pp. 384-390.

Lieberman, S.H., P. Boss, G.W. Anderson, G. Heron, and K.S. Udell, 2000 (in print). "Characterization of NAPL Distributions Using In-Situ Imaging and LIF", in *Proceedings for the Second International Conference on Remediation of Chlorinated and Recalcitrant Compounds*, May 22-25, 2000, Monterey, CA.

Looney, Brian B., Karen M. Jerome, and Christine Davey, 1998. "Single Well DNAPL Characterization Using Alcohol Injection/Extraction", in *Proceedings for the First International Conference on Remediation of Chlorinated and Recalcitrant Compounds, Non-Aqueous-Phase Liquids*, Monterey, CA, May 18-21, 1998, pp. 113-118.

Marrin, Donn L., 1988. "Soil-gas Sampling and Misinterpretation", *Ground Water Monitoring Review*, v. 8, pp. 51-54.

Marrin, Donn L. and Henry B. Kerfoot, 1988. "Soil-gas Surveying Techniques", *Environmental Science and Technology*, v. 22, pp. 740-745.

Meinardus, Hans W., Richard E. Jackson, Minquan Jin, John T. Londergan, Sam Taffinder, and John S. Ginn, 1998. "Characterization of a DNAPL Zone with Partitioning Interwell Tracer Tests", in *Proceedings for the First International Conference on Remediation of Chlorinated and Recalcitrant Compounds, Non-Aqueous-Phase Liquids*, Monterey, CA, May 18-21, 1998, pp. 143-148.

Mosier-Boss, Pamela A., Ricardo Newbery, and Stephen H. Lieberman, 1997. "Development of a Cone Penetrometer Deployed Solvent Sensor Using a SERS Fiber Optic Probe", in *Proceedings for a Specialty Conference on Field Analytical Methods for Hazardous Wastes and Toxic Chemicals*, January 29-31, 1997, Las Vegas Nevada, pp. 588-599.

MSE Technology Applications, Inc., 2000 (*in print*). "Cost Analysis of Dense Non Aqueous Phase Liquid Characterization Tools", prepared for U.S. Department of Energy, Contract No. DE-AC22-96EW96405.

Nelson, N.T. and M.L. Brusseau, 1996. "Field Study of the Partitioning Tracer Method for Detection of Dense Nonaqueous Phase Liquid in a Trichloroethene-Contaminated Aquifer", *Environmental Science and Technology*, v.30, no. 9, pp. 2859-2863.

Nelson, N.T., M. Oostrom, T.W. Wietsma, and M.L. Brusseau, 1999. "Partitioning Tracer Method for the In Situ Measurement of DNAPL Saturation: Influence of Heterogeneity and Sampling Method", *Environmental Science and Technology*, v.33, p.4046-4053.

Newell, C. and R.R. Ross, 1991. "Estimating Potential for Occurrence of DNAPL at Superfund Sites, Quick Reference Guide Sheet", U.S. Environmental Protection Agency, publication number 9355.4-07FS, Washington, D.C.

Pankow, James F., and John A. Cherry, 1996. *Dense Chlorinated Solvents and other DNAPLs in Groundwater: History, Behavior, and Remediation*, Waterloo Press, Portland, Oregon, 522 pages.

Payne, Tamra, James Brannon, Ronald Falta, and Joseph Rossabi, 1998. "Detection Limit Effects on Interpretation of NAPL Partitioning Tracer Tests", in *Proceedings for the First International Conference on Remediation of Chlorinated and Recalcitrant Compounds, Non-Aqueous-Phase Liquids*, Monterey, CA, May 18-21, 1998, pp. 125-130.

Pitkin, Seth E., 1998. "Detailed Subsurface Characterization Using the Waterloo Profiler", in Proceedings for the Symposium on the Application of Geophysics to Environmental and Engineering Problems, March 22-26, 1998, Chicago, Ill., pp. 53-64.

Rossabi, J., B.D. Riha, C.A. Eddy-Dilek, A. Lustig, M. Carrabba, W.K. Hyde, and J. Bello, 2000 (in print). "Field Tests of a DNAPL Characterization System Using Cone Penetrometer-based Raman Spectroscopy", <u>Ground Water Monitoring and Remediation</u>, 2000, in print.

Semprini, Lewis, Michael Cantaloub, Sarayu Gottipati, Omar Hopkins, and Jonathan Istok, 1998. "Radon-222 as a Tracer for Quantifying and Monitoring NAPL Remediation", in *Proceedings for the First International Conference on Remediation of Chlorinated and Recalcitrant Compounds, Non-Aqueous-Phase Liquids*, Monterey, CA, May 18-21, 1998, pp. 137-142.

Sinclair, N. and M. Kram, 1998. "High Resolution 3-D Seismic Reflection Surveys for Characterization of Hazardous Waste Sites", in *Proceedings for the Third Tri-Service ESTCP Workshop*, 18 August, 1998.

Sudicky, E.A., 1986. "A Natural Gradient Experiment on Solute Transport in a Sand Aquifer: Spatial Variability of Hydraulic Conductivity and Its Role in the Dispersion Process", *Water Resources Research*, v. 22, no. 13, pp. 2069-2082.

Taylor, T.W. and M.C. Serafini, 1988. "Screened Auger Sampling: The Technique and Two Case Studies", *Ground Water Monitoring Review*, v. 8, no. 3, pp. 145-152.

Wise, W.R., 1999. "NAPL Characterization via Partitioning Tracer Tests: Quantifying Effects of Partitioning Nonlinearities", *Journal of Contaminant Hydrology*, vol. 36, no. 1-2, pp. 167-183.

Yoon, S., I. Barman, A. Datta-Gupta, and G.A. Pope, 1999. "In-Situ Characterization of Residual NAPL Distribution Using Streamline-Based Inversion of Partitioning Tracer Tests", in *Proceedings: The 1999 Exploration and Production Environmental Conference, SPE/EPA*, Austin, TX, USA, 03/01-03/99, pp. 391-400.

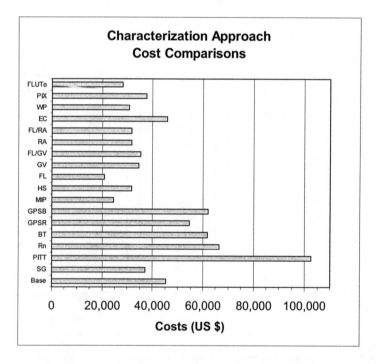

FIGURE 1. Costs for each DNAPL Characterization Approach. [Base = Baseline; SG = Soil Gas Survey; PITT = Partition Interwell Tracer Test; Rn = Radon Flux Survey; BT = Solute Back-Track; GPSR = Surface Geophysical; GPSB = Subsurface Geophysical; MIP = membrane Interface Probe; HS = Hydrosparge; FL = Fluorescence Probe; GV = GeoVis Probe; FL/GV = Fluorescence Probe with GeoVis; RA = Raman Probe; FL/RA = Fluorescence with Raman Probe; EC = Electrochemical Sensor Probe; WP = Waterloo Profiler; PIX = Precision Injection-Extraction; FLUTe = Flexible Liner Underground Everting Membrane].

HELIUM AND NEON AS DNAPL PARTITIONING TRACERS

Craig E. Divine (ARCADIS Geraghty & Miller, Denver Colorado, USA)
William E. Sanford (Colorado State University, Fort Collins, Colorado, USA)

ABSTRACT: Recent work has demonstrated that the partitioning interwell tracer test (PITT) can be used to locate and quantify subsurface dense nonaqueous phase liquid (DNAPL). Laboratory investigation indicates that dissolved helium and neon are appropriate partitioning tracer candidates for field-scale PITT application. Batch experiments determined the trichloroethene (TCE)-water equilibrium partition coefficients to be 2.40 ±0.18 for helium and 3.36 ±0.22 for neon. Tracer breakthrough curves (BTCs) from column tests were analyzed by direct integration and by fitting a dual-porosity transport model to the data, yielding similar average error in calculated TCE saturation (11% and 13%, respectively). Sensitivity analysis shows that low tracer detection limits are more important than tracer measurement precision, and that accurate characterization of the tail region of the BTC is particularly important.

INTRODUCTION

Successful and cost-efficient groundwater remediation depends upon accurate characterization of contaminant mass and distribution, including source areas (i.e., DNAPL zones). However, traditional site characterization methods, such as drill-core analysis and site history review, rarely provide enough information to reliably characterize DNAPL distribution and saturation (Pankow and Cherry, 1996). One promising new technique for characterizing subsurface nonaqueous phase liquid (NAPL) is the PITT, which was originally developed in the early 1970s by researchers in the oil industry (see Zemel, 1995).

Tracer BTC data from a PITT is analyzed to estimate the average pore NAPL saturation (S_N) within the region swept by the tracers. By definition, the transport of a conservative tracer is unaffected by the presence of residual NAPL, while partitioning tracers will be retarded by NAPL. The magnitude of tracer retardation depends upon S_N and the tracer-specific NAPL-water partition coefficient (K), which is given by:

$$K = \frac{c_N}{c_W} \tag{1}$$

where c_N is the tracer concentration in NAPL and c_W is the tracer concentration in water. The tracer-specific retardation coefficient (R) is the average tracer velocity divided by the groundwater velocity. These values are used to calculate S_N by:

$$S_N = \frac{R-1}{R+K-1} \tag{2}$$

Several successful field-scale PITTs have been reported in the literature (see Jin et al., 1995; Jin 1995; Wilson and Mackay, 1995; and Nelson and Brusseau, 1996). Alcohols have been used as partitioning tracers for the majority of PITTs conducted in the saturated zone; however, a disadvantage with alcohol tracers is their relatively high analytical detection limits. For example, Young et al. (1999) report on a PITT where the range from the tracer source concentration to the detection limit was approximately three orders of magnitude. This analytical limitation can prevent accurate characterization of the tracer BTC, particularly in the tail region of the curve. Payne et al. (1998) demonstrate that NAPL present within low permeability zones may go undetected when high tracer detection limits prevent accurate BTC characterization.

Helium and neon have low atmospheric concentrations, low detection limits, and can be easily measured on a simple gas chromatograph (GC). The range from measurable background concentrations to solubility for these gases can be as large as five orders of magnitude, which may permit greater accuracy in BTC characterization. Helium and neon are nontoxic, chemically stable, and are relatively inexpensive. Examples of dissolved gas groundwater tracer tests can be found in Carter et al. (1959), Sugisak (1961), Gupta et al. (1994), and Sanford and Solomon (1998). Sanford et al., (1996) developed simplified methods for dissolved gas application, including a diffusive *in situ* dissolved gas sampler constructed of permeable silicone tubing and a Schrader-type tire stem valve (shown in Figure 1). This passive sampler can easily collect a variety of dissolved gases including: helium, neon, H_2, Ar, CH_4, N_2, Kr, and SF_6. Sanford and Solomon (1998) used these methods to conduct a relatively inexpensive year-long dissolved gas tracer test. The advantages of dissolved gas tracers motivated this laboratory investigation of the applicability of helium and neon as partitioning tracers, and the results provide a basis for their application in field-scale PITTs.

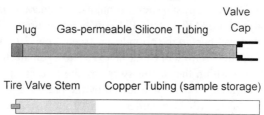

FIGURE 1. Diffusive sampler described by Sanford et al. (1996).

EXPERIMENTAL METHODS

Samples were analyzed for bromide with an Orion® ion-selective electrode and a single junction reference electrode. Water samples were analyzed for dissolved helium and neon by direct headspace analysis with a bench-top GC equipped with a thermal conductivity detector and nitrogen carrier gas. The lower practical quantitation limit (PQL) was approximately 7×10^{-4} cm³ L⁻¹ for helium and 5×10^{-3} cm³ L⁻¹ for neon. Natural background concentrations of helium and neon were detected in water and it is probable that a much lower PQL could be achieved with slight modifications to the GC specifications (reducing the flow rate and using argon carrier gas, for example). Natural background concentrations in water are approximately 4.5×10^{-5} cm^3 L^{-1} for helium and 1.8×10^{-4} cm^3 L^{-1} for neon (Sanford et al., 1996).

The TCE-water partition coefficients for helium and neon were determined from a series of batch partitioning tests conducted in 10-cm^3 glass syringes. After equilibrium was achieved within the syringe, helium and neon concentrations were measured in the TCE and water fractions, and the TCE-water partition coefficients were calculated by Equation 1.

Column partitioning tracer experiments were conducted in 30-cm long, 5-cm diameter, glass liquid chromatography columns. The columns were packed with a well-sorted medium Ottawa sand standard under saturated conditions. During column construction and immediately prior to tracer tests, the column was flushed with de-aired water to prevent the entrapment of air bubbles and ensure complete saturation. TCE dyed with SUDAN IV was introduced into the water-saturated column with a syringe, and physically distributed to prevent the formation of continuous NAPL. Columns were constructed with TCE saturation ranging from 0.047 to 0.105 (to minimize the possibility of continuous NAPL, experiments were not conducted with TCE saturation greater than 0.105). The columns were oriented vertically and upward water flow was maintained with peristaltic pump. A sketch of the column set-up is provided in Figure 2.

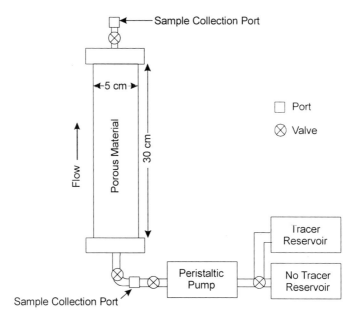

FIGURE 2. Sketch of column test set up.

Temporal moment analysis of BTC data can provide average tracer travel times; however, temporal moments may only be estimated for finite data distributions, such as BTCs.   Error in moment estimation is increased by measurement noise, low sampling frequency, and BTC truncation.   Temporal moments were estimated directly from the data (Direct Integration method) and from infinite data distributions generated from solute transport models fitted to the experimental BTC (Model Fitting method).

Direct Integration is the traditional method to estimate temporal moments, and the advantage with this method is that the BTC data is evaluated directly without assumptions regarding the solute transport model. However, if there are data gaps, noise due to experimental error, or poor characterization of the BTC tail due to analytical limitations or early test cut-off, error may be significant. Model Fitting minimizes the effects of noise and data gaps in the observed BTC and can be used to estimate other important solute transport parameters; however, an appropriate solute transport model must be chosen.   A one-dimensional advection-dispersion equilibrium model (see van Genuchten, 1981) was fitted to the conservative tracer (bromide) data, and to the helium and neon BTCs when $S_{TCE} = 0$. The transport of the partitioning tracer was modeled using a physical two-region (dual porosity) nonequilibrium solute transport model. This model assumes that there are two distinct liquid regions: mobile (water) and immobile (NAPL) with diffusion controlled solute transfer between the two regions (see van Genuchten and Wagenet, 1989). The unknown parameters in the transport models were estimated with the fitting computer program CXTFIT 2.1.

Note that temporal moment analysis is not appropriate for non-linear processes (Valocchi, 1985) which may exist if there is significant diffusion within large continuous NAPL zones (i.e., pools and fingers). Based upon the results of the column experiments, temporal moment analysis is acceptable when NAPL distribution is discontinuous, or residual.   Divine (2000) provides further discussion of temporal moment estimation techniques for these experiments.

RESULTS AND DISCUSSION

Based on the results of the batch tests, the TCE-water partition coefficients are 2.40 ±0.18 for helium and 3.36 ±0.22 for neon (95% confidence). The results of eight column tests are shown in Table 1. Superimposed partitioning tracer BTCs from two tests conducted on the same column are shown in Figure 3 (the bromide BTCs for the two tests were nearly identical, agreeing within 2%).

TABLE 1. Summary of results for column partitioning tracer experiments.

Actual S_{TCE}	Partitioning Tracer	Direct Integration		Model Fitting	
		Estimated S_{TCE}	Error	Estimated S_{TCE}	Error
0.047	Neon	0.039	-0.008 (-17%)	0.040	-0.007 (-14.9%)
0.052	Helium	0.054	0.002 (3.8%)	0.070	0.018 (34.6%)
0.083	Helium	0.086	0.003 (3.6%)	0.112	0.029 (34.9%)
0.091	Helium	0.102	0.011 (12.1%)	0.091	0.000 (0.0%)
0.096	Helium	0.074	-0.022 (-22.9%)	0.087	-0.009 (-9.4%)
0.097	Helium	0.079	-0.018 (-18.6%)	0.107	0.010 (10.3%)
0.097	Neon	0.104	0.007 (7.2%)	0.097	0.000 (0.0%)
0.105	Helium	0.107	0.002 (1.9%)	0.108	0.003 (2.9%)

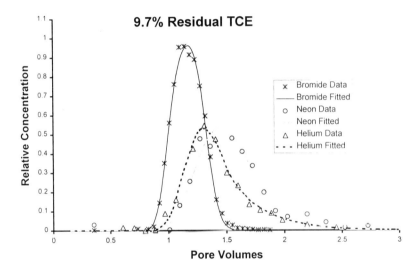

FIGURE 3. Superimposed BTCs for a column with 9.7% residual TCE.

The average S_{TCE} estimation error was similar for both methods, with an absolute error of approximately 0.009 (11%) by Direct Integration and 0.010 (13%) by Model Fitting. A sensitivity analysis was performed to further understand S_N estimation error. The dual-porosity model was used to create a BTC with predetermined solute transport values. A series of synthetic BTCs were then generated by introducing error representing tracer measurement error, low sampling frequency, and high tracer detection limits. Model Fitting tended to overestimated S_N while Direct Integration underestimated S_N, particularly when the BTC tail was truncated or poorly characterized. Random tracer measurement error is relatively unimportant, but high sampling frequency should be maintained until tracer concentrations fall below the detection limit. Furthermore, tracers with low detection limits will allow better BTC characterization and minimize the error associated with BTC truncation.

The PITT technique is appropriate for measurement of both DNAPL and light nonaqueous phase liquid (LNAPL); however, due to their high Henry's coefficients, dissolved helium and neon will be retarded by trapped air. Trapped air may be the result of seasonal water level changes, variable pumping rates from extraction wells, or air sparging. Since LNAPLs are typically encountered near the water table where the presence of trapped air is likely, helium and neon may not be suitable for LNAPL quantification. However, dissolved helium and neon could be used to measure trapped air by methods similar to the PITT. Additionally, helium and neon are good candidates as conservative tracers for water saturation estimation in the unsaturated zone by a modified PITT technique.

CONCLUSION

Laboratory investigation indicates that dissolved helium and neon are appropriate partitioning tracer candidates for field-scale PITT application. Batch experiments determined the TCE-water equilibrium partition coefficients to be 2.40 ±0.18 for helium and 3.36 ±0.22 for neon. Column-scale partitioning tracer experiments were conducted with S_{TCE} values ranging from 0.047 to 0.105. The average S_{TCE} estimation error was approximately 11% by the Direct Integration technique and 13% by the Model Fitting technique. Sensitivity analysis shows that low tracer detction limits are more important than measurement precision, and that accurate characterization of the tail region of the BTC is particularly important. Due to their high dimensionless Henry's coefficients, dissolved helium and neon will be retarded by the presence of trapped air.

REFERENCES

Carter, R. C., W. J. Kaufman, G. T. Orlob, and D. K. Todd. 1959. "Helium as a Ground-water Tracer." *Journal of Geophysical Research.* 64(12): 2433-2439.

Divine, C. E. 2000. "The Applicability of Dissolved Helium and Neon as Dense Nonaqueous Phase Liquid (DNAPL) Partitioning Tracers." M.S. Thesis, Colorado State University, Fort Collins, CO.

Gupta, S. K., P. S. Moravcik, and L. S. Lau. 1994. "Use of Injected Helium as a Hydrological Tracer." *Hydrological Sciences Journal.* 39: 109-119.

Jin, M. 1995. "A Study of Nonaqueous Phase Liquid Characterization and Surfactant Remediation." Ph.D. Dissertation, University of Texas, Austin, TX.

Jin, M., M. Delshad, V. Dwarakanath, D. C. McKinney, G. A. Pope, K. Sepehrnoori, C. E. Tilberg, and R. E. Jackson. 1995. "Partitioning Tracer Test for Detection, Estimation, and Remediation Performance Assessment of Subsurface Nonaqueous Phase Liquids." *Water Resources Research.* 31: 1201-1211.

Nelson, N. T., and M. L. Brusseau. 1996. "Field Study of the Partitioning Tracer Method for Detection of Dense Nonaqueous Phase Liquid in a Trichloroethene Contaminated Aquifer." *Environmental Science and Technology.* 30: 2895-2863.

Pankow, J. F. and J. A. Cherry. 1996. *Dense Chlorinated Solvents and Other DNAPLs in Groundwater.* Waterloo Press, Portland, OR.

Payne, T., J. Brannan, R. Falta, and J. Rossabi. 1998. "Detection Limit Effects on Interpretations of NAPL Partitioning Tracer Tests." In *Proceedings of 1st International Conference on Remediation of Chlorinated and Recalcitrant Compounds,* pp.0125-130. Monterey, CA.

Sanford, W. E., R. G. Shropshire, and D. K. Solomon. 1996. "Dissolved Gas Tracers in Groundwater: Simplified Injection, Sampling, and Analysis." *Water Resources Research.* 32(6): 1635-1642.

Sanford, W. E. and D. K. Solomon. 1998. "Site Characterization and Containment Assessment with Dissolved Gases." *Journal of Environ. Eng.* 124 (6): 572-574.

Sugisaki, R. 1961. "Measurement of Effective Flow Velocity of Ground Water by Means of Dissolved Gases." *American Journal of Science.* 259: 144-153.

Valocchi, A. J. 1985. "Validity of the Local Equilibrium Assumption for Modeling Sorbing Solute Transport Through Homogeneous Soils." *Water Resources Research.* 21(6): 808-820.

van Genuchten, M. Th. 1981. "Analytical Solutions for Chemical Transport With Simultaneous Adsorbtion, Zero Order Production, and First Order Decay." *Journal of Hydrology.* 49: 213-233.

van Genuchten, M. Th. and R. J. Wagenet. 1989. "Two-site/two-region Models for Pesticide Transport and Degradation: Theoretical Development and Analytical Solutions." *Soil Science Society of America Journal.* 53: 1303-1310.

Wilson, R. D. and D. M. Mackay. 1995. "Direct Detection of Residual Nonaqueous Phase Liquid in the Saturated Zone using SF_6 as a Partitioning Tracer." *Environmental Science and Technology.* 29: 1255-1258.

Young, C. M., R. E. Jackson, M. Jin, J. T. Londergan, P. E. Mariner, G. A. Pope, F. J. Anderson, and T. Houk. 1999. "Characterization of a TCE DNAPL Zone in Alluvium by Partitioning Tracers." *Ground Water Mon. and Remed.* 19(1): 84-94.

Zemel, B. 1995. *Tracers in the Oil Field.* Elsevier Science B.V. Amsterdam.

SIMULATION OF SURFACTANT-ENHANCED PCE RECOVERY AT A PILOT TEST FIELD SITE

Chad D. Drummond, Lawrence D. Lemke, Klaus M. Rathfelder, Ernest J. Hahn, and Linda M. Abriola, The University of Michigan, Ann Arbor, MI USA

ABSTRACT: Characterization studies of a tetrachloroethylene (PCE) contaminated site at a former dry cleaning facility suggest that it is a good candidate for the application of a pilot scale surfactant-enhanced aquifer remediation (SEAR) field demonstration project. Soil core samples have confirmed the presence of residual PCE within a coarse sand and gravel layer near the water table and above the underlying clay layer. A SEAR pilot test was designed with three-dimensional flow and transport models for evaluation of the hydraulic parameters, and laboratory studies were conducted for surfactant screening and selection. The SEAR design was further evaluated with a two-dimensional enhanced solubilization simulator. The potential influence of small-scale heterogeneity on organic distribution, surfactant delivery, and solution recovery was assessed with geostatistical models of the formation heterogeneity conditioned to grain size distribution analyses performed on soil cores collected from the site. Simulation results suggest that SEAR can effectively remove PCE from this site, provided that a significant contaminant volume is not sequestered in a sand/silt/clay transition zone located near the bottom of the aquifer. In addition, simulation results and predicted recovery rates may be influenced by uncertainties in permeability distribution, flow field, and PCE distribution.

INTRODUCTION

Aquifers contaminated with non-aqueous phase liquids (NAPLs) present some of the most difficult and costly remediation challenges facing hydrogeologists and engineers. Traditional pump-and-treat methods are generally considered to be inefficient for NAPL recovery due to their inability to mobilize entrapped NAPLs and the generally low solubility of organic liquids. The use of surfactants to increase organic liquid solubility and/or increase NAPL mobility is a promising alternative remediation strategy. SEAR has been demonstrated to be highly effective at removing NAPLs in a number of laboratory column and sand box experiments (e.g. Pennell et al., 1993; Taylor et al., 2000).

In field settings, the effectiveness of SEAR will be influenced by soil heterogeneities. Taylor (1999) reviewed 23 field trials of SEAR technology for source zone remediation conducted through 1998. The SEAR demonstrations exhibited mixed performance ranging from little or no mass recovery to mass recoveries up to 75-99%. In general, SEAR often failed at sites where no attempt was made to characterize the site subsurface or where laboratory evaluation of site contaminants and soils was omitted. SEAR was most successful at sites that were relatively homogeneous and underlain by a confining clay layer. Heterogeneities generally diminished the effectiveness of SEAR by impeding surfactant delivery to the source zone region.

This paper describes the design of a SEAR pilot demonstration project at a former dry cleaning facility in Oscoda, Michigan. The design of the pilot test incorporates 1) site characterization studies for source zone delineation and description of soil heterogeneities, 2) laboratory studies for surfactant evaluation and performance, and 3) the use of numerical models and geostatistical tools for system design assessment.

SITE DESCRIPTION AND CHARACTERIZATION

The SEAR demonstration test site is located beneath a former dry cleaning facility in Oscoda, Michigan (Figure 1) referred to as the Bachman Road site. The aquifer is unconfined with a saturated thickness of approximately 15 feet (4.6 m), and comprised of a fairly uniform fine to medium sand. A relatively flat basal clay layer is present at a depth of 24 feet (7.3 m). A transitional zone in which the grain size decreases from that of sand to silt to clay exists in the region 1-2 feet (0.3-0.6 m) above the clay. The medium and fine sands have a low organic content and low sorptive capacity, whereas the silt and clay have a high sorptive capacity. The primary contaminant at the site is PCE, which is suspected to have been released near the back half of the building (shaded area in Figure 1). The volume of PCE released is unknown. A narrow PCE plume emanates from the source area and discharges into Lake Huron approximately 700 feet (213 m) down-gradient. Plume centerline aqueous PCE concentration measurements range from a high of 88 ppm beneath the source zone to 10 ppm in observation wells 500 feet (152 m) down-gradient.

A Geoprobe drivepoint system was used to collect aqueous samples and soil cores at varying depths (Figure 1). Additionally, five bundled multi-level monitoring wells with a total of 26 sampling ports (not shown) were installed in the source region. These wells have been sampled on two occasions to establish baseline conditions and will be sampled periodically during the SEAR test. The greatest aqueous PCE concentrations have been detected along a narrow band emanating from the source region, stretching down-gradient in the direction of the regional flow. In this region, consistently high PCE concentrations (up to 25 ppm) were detected near the bottom of the aquifer. Very high concentrations were also detected at depths between 14-16 ft (4.3–4.9 m) (up to 50 ppm in the front parking lot and up to 88 ppm behind the building). Soil cores analyses similarly showed consistently large concentrations above the clay and between depths 11-16 ft (3.4-4.9 m). PCE concentrations in soil ranged up to 4,800 ppm in the upper portions of the aquifer, confirming the presence of residual PCE at several locations in the suspected source region. Soil concentrations above the clay were consistently elevated, but could not be used to confirm the presence of residual NAPL due to the high sorptive capacity of the silt and clay. Because of the limitations of angle coring, soil samples above the clay were mostly obtained near the perimeter of the building and could not be obtained under the center of the building.

Sediment beneath the Bachman Road site consists predominantly of fine to medium grained sand; however, variability in grain size is discernable in continuous core samples taken at the site. Grain size distributions (GSDs) were measured for 167 core samples taken at 0.5 to 2.0 foot (0.15 to 0.61 m) intervals along eight vertical and angled borings at the site. To remove any bias in the mean grain size estimate created by gravel size material, the GSDs were renormalized to 100% after excluding the sediment fraction coarser than 850 μm. Hydraulic conductivity was then estimated with the Carman-Kozeny relationship. Estimated hydraulic conductivities ranged between 15-150 ft/d (4.6-45.7 m/d). A high permeability, coarse sand and gravel layer was consistently detected at depths between 12-16 ft (3.7-4.9 m).

Characterization of the Bachman Road site indicates that it has several characteristics that Taylor (1999) associated with successful and effective SEAR applications. These factors include a fairly homogeneous aquifer underlain by a confining clay layer and the detection of contaminants in a coarse sand layer. Based on these studies, the Bachman Road site appears to be a good candidate for a pilot scale SEAR test, targeting source zone remediation.

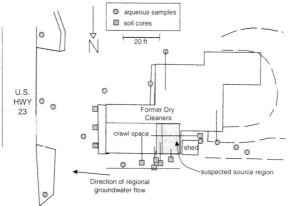

FIGURE 1. Site plan and Geoprobe locations.

PILOT TEST DESIGN

Surfactant Selection and Evaluation. Three surfactants were evaluated for use during the pilot test: 1) Witconol 2722, a food grade surfactant that has a high capacity to solubilize PCE, and exhibits little macroemulsion formation and should not displace residual PCE from Bachman aquifer material; 2) Aerosol MA/OT, a food grade surfactant which has a high capacity to reduce interfacial tension, resulting in the displacement and mobilization of residual PCE; and 3) a combination of Aerosol MA, isopropanol, and salts which tend to increase solubilization and lower the interfacial tension.

The surfactants were experimentally evaluated by measuring their ability to flush PCE from Bachman aquifer materials in both rectangular columns and in two-dimensional sand box systems containing macro-heterogeneities. The

Aerosol MA/OT solubilized and displaced PCE but was inefficient due to undesirable density override behavior.   The Aerosol MA/isopropanol/salt formulation solubilized and mobilized the residual PCE, which in the two-dimensional sand tank system resulted in the downward migration of PCE into a fine grained confining layer.   Witconol 2722 produced solubilization recovery rates comparable to the Aerosol MA/isopropanol/salt formulation without any observable PCE migration.   Thus, to avoid mobilizing PCE into the transitional zone at the Bachman site, Witconol 2722 was selected for the Bachman Road pilot study.

Hydraulic Design And Flow Field Prediction.   Numerical simulations of the pilot test surfactant delivery/removal system were performed using industry standard software.   VMODFLOW v2.8.2 was used to predict the natural and engineered flow fields, MT3DMS was employed to predict surfactant transport, and MODPATH was used to predict advective surfactant sweep using particle tracking.   Initial flow modeling consisted of a regional-scale model that was calibrated using available monitoring well head data.   Telescoping was then performed to obtain a local model of the SEAR test site.   The local model employed ten horizontal layers of differing hydraulic conductivity ranging from 23 to 99 ft/d (30.2 to 7.0 m/d) determined from grain size distributions.

The primary design parameters for the pilot test are well locations and pumping schedules.   These parameters were varied to maximize delivery of surfactant to the source zone, minimize injected surfactant volume, and prevent surfactant losses outside the source zone.   The final pilot test design consists of a single extraction well (5.2 gal/min, 19.7 L/min), a row of three water injection wells (1.0 gal/min, 3.8 L/min each), used to establish a flow field through the source zone, and a second row of three surfactant injection wells (0.5 gal/min, 1.9 L/min each) positioned between the water supply and extraction wells (Figure 2). Two pore volumes of a Witconol 2722 surfactant solution will be injected at a concentration of 40 g/L (4%) through the surfactant injection wells over a ten day period.   Table 1 lists design parameters used for these simulations and Figure 2 shows predicted pathlines at a depth of 3 ft (0.9 m) below the water table. Pathlines predict that the well arrangement will effectively sweep the entire pilot test area while minimizing surfactant travel outside of the test zone.   The sharp change in pathline direction is due to a transient in pumping at day 16.

Table 1. SEAR Pilot Test Parameters.

	Surfactant Injection (S1, S2, S3)	Water Injection (W1, W2, W3)	Extraction (E1)
Pumping Rate (gpm)	0.5	1.0	5.2
Days of Operation	2 – 16 Days 2 – 6, water only	2 - 16	1 - 45
Days Injecting Surfactant	7 - 16	--	--

DESIGN EVALUATION
        Numerical simulation models were used to further evaluate the potential effectiveness of the proposed SEAR pilot project.   To the extent possible,

information about the site geology, PCE migration and entrapment behavior, and surfactant enhanced solubilization processes were incorporated in the numerical evaluation. This necessitated the use of a series of models: 1) geostatistical modeling for the estimation of permeability distribution; 2) multiphase flow simulations for estimation of PCE distribution; and 3) simulation of enhanced solubilization processes for evaluation of PCE recovery.

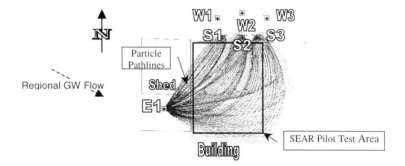

FIGURE 2. Pilot test site layout.

Geostatistical Heterogeneity Modeling. To develop realizations of the conductivity distribution, experimental semivariograms were constructed using the normal score transform of estimated hydraulic conductivity values (Deutsch and Journel, 1998). Vertical and omnidirectional horizontal semivariograms were modeled using a zonal anisotropy model combining two spherical models with a nugget effect of 0.235. In the vertical direction, the actual range was 2.5 feet (0.76 m) with a standardized sill (including the nugget effect) of 1.0. In the horizontal direction, the actual range was 20.0 feet (6.1 m) with a standardized sill of 0.823. A three-dimensional realization of conductivity values, conditioned to the original 167 estimates, was generated at a one-foot grid increment across the study area using a sequential Gaussian simulation algorithm (Deutsch and Journel, 1998). A 22x16 foot (6.7x4.9 m) XZ profile (vertical east-west) was then extracted from the three-dimensional conductivity field. Simulated conductivity values along the profile ranged from 8.8 to 118.4 ft/d (2.7 to 36.1 m/d) (Figure 3).

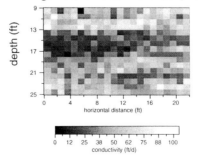

Figure 3. XZ cross section of hydraulic conductivity.

Prediction of Initial PCE Distribution. The distribution of PCE will have a major influence on SEAR performance. Unfortunately, the volume of PCE is unknown and information about the PCE distribution is very limited. Aquifer heterogeneity can have a large impact on organic contaminant distribution (Dekker and Abriola, 2000). To develop estimates of PCE distribution, a two-dimensional simulator, M-VALOR (Abriola et al., 1992), was used to predict possible initial PCE distributions in the formation. PCE infiltration at the Bachman site was simulated using two distributed sources representing suspected source regions; the first near the back door of the building, and the second through the floor boards of the building. A uniform release rate of 60 mL/d for a period of 400 days was used. Redistribution was simulated for a total of 5 years. Distributions were predicted for two cases: A) an average homogeneous permeability distribution (59 ft/d, 18 m/d) and B) the heterogeneous permeability distribution shown in Figure 3. In both cases a one-foot vertical transition zone with a uniform conductivity of 7 ft/d (2.1 m/d) was included at the bottom of the domain. Simulations were performed on a uniform one-half foot by one-half foot grid, and entry pressures were estimated with a Leverett scaling procedure.

Figure 4 shows predicted PCE distributions. The predicted PCE distribution in the homogeneous case exhibits a regular downward migration path. The downward migration and head build-up was sufficient to overcome the entry pressure of the transition zone, resulting in pooling on top of the confining layer within the transition zone. In contrast, the predicted PCE distribution in the heterogeneous system exhibits an irregular downward migration path with pooling on top of the transition layer and no penetration into the transition region. Note that identical grid spacing and soil properties of the transition zone were used in both simulations. Further note that for plotting purposes the saturations in Figure 4 are limited to 3%, whereas predicted saturations ranged up to 5% in the homogeneous case and up to 10% in the heterogeneous case. Both distributions show a large proportion of the mass collects in pools near the aquifer bottom.

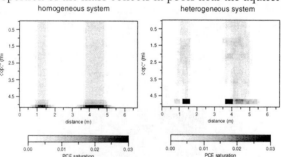

Figure 4. Predicted PCE distributions

Two-Dimensional Multiphase Solubilization Simulations. SEAR at the Bachman Road site was simulated in the two-dimensional cross sectional domain, roughly tracing a flow path between the center surfactant injection well and the extraction well (Figure 2). Flow rates and surfactant concentrations in the injection well were adjusted to obtain a qualitative match with breakthrough profiles and concentrations predicted in the three-dimensional flow and transport code used for the hydraulic design. Surfactant solution was injected at a rate of 0.167 gpm/ft along the vertical boundary, and was apportioned by conductivity. Surfactant was injected at a concentration of 3% (wt) for a period of 10 days, followed by injection of clean water for another 10 days to flush the surfactant. The SEAR application was simulated with a two-dimensional solubilization simulator, MISER (Abriola et al. 1997). This model accounts for rate limited PCE solubilization and surfactant sorption. Solubilization mass transfer rates were evaluated from correlations determined from column flushing experiments (Taylor et al., 2000). The enhanced solubilization simulator has been previously validated with data from SEAR experiments conducted in laboratory sand tank systems (Rathfelder et al., 2000).

Figure 5 shows predicted PCE and surfactant breakthrough at the extraction well for the homogeneous and heterogeneous permeability distributions. Greater breakthrough concentrations are predicted in the homogeneous system. However concentrations fall off more rapidly and exhibit tailing behavior due to the sequestering of NAPL in the transition zone. Breakthrough in the heterogeneous systems occurs earlier due to the presence of preferential flow paths. Also PCE recovery is greater and exhibits less tailing behavior. Density plunging of the surfactant solutions did not appear to influence PCE recovery in these simulations. Surfactant breakthrough behavior was similar to that obtained for PCE. Recovery of surfactant along this cross section was about 97% after 20 days, with the remaining surfactant primarily contained on the solid phase.

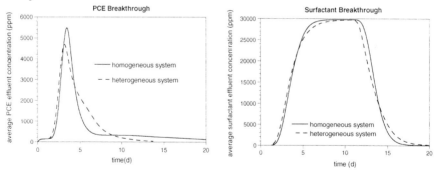

Figure 5. Predicted PCE and surfactant breakthrough at the extraction well.

SUMMARY AND DISCUSSION

Conventional groundwater flow and transport models were used to develop a hydraulic design for a surfactant enhanced aquifer remediation pilot test at a former dry cleaning facility. Because of uncertainty in the permeability and PCE distribution at the field site, a two-dimensional solubilization simulator was used to assess the potential effectiveness for PCE removal from the suspected source region. Aquifer heterogeneity was incorporated into the modeling study using geostatistical information obtained from soil core data. A significant result imparted by the heterogeneities was that PCE was predicted to pool on top of the transition zone in the heterogeneous system, whereas PCE penetrated and pooled within the transition layer in the homogeneous system. The presence of PCE within the transition zone resulted in lower predicted PCE recovery and pronounced tailing behavior. Model predictions generally suggest that PCE can be effectively removed from this site, and that the removal efficiency will depend to a large extent on the proportion of contaminants within the transition zone. The presence of residual PCE within the transition zone has not been confirmed due to the high sorptive capacity of silt and clay, and the inability to obtain cores at depth in the center of the building. It is also emphasized that the numerical results are exploratory in nature, and that uncertainties in the permeability distribution, flow field, and PCE distribution can impact predicted recovery rates.

ACKNOWLEDGEMENTS
The Michigan Department of Environmental Quality under Contract No. Y80011 and the U.S. Environmental Protection Agency, Great Lakes and Mid-Atlantic Center for Hazardous Substance Research (GLMAC-HSRC) under Grant No. R-825540 provided funding for this research.

REFERENCES
Abriola, L.M., K.M. Rathfelder, S. Yadav & M. Maiza. 1992. *A PC Code for Simulating Subsurface Immiscible Contaminant Transport.* EPRI TR-101018, Electric Power Research Institute, Palo Alto.

Abriola, L.M., J.R. Lang & K.M. Rathfelder. 1997. Michigan Soil Vapor Extraction Remediation (MISER) Model: A Computer Program to Model Soil Vapor Extraction and Bioventing of Organic Chemicals in Unsaturated Geologic Materials, U.S. Environmental Protection Agency, National Risk Management Research Laboratory, Ada OK.

Dekker, T. J., and L. M. Abriola. "The Influence of Field Scale Heterogeneity on the Infiltration and Entrapment of Dense Nonaqueous Phase Liquids in Saturated Formations." *J. Cont. Hydrol.*, in press-a, 2000.

Deutsch, C. V. and A. G. Journel. 1998. *GSLIB Geostatistical Software Library and User's Guide: Applied Geostatistics Series.* New York, Oxford University Press.

Pennell, K. D., G. A. Pope, L. M. Abriola. 1996. "Influence of Viscous and Buoyancy Forces on the Mobilization of Residual Tetrachloroethylene during Surfactant Flushing." *Environ. Sci. Technol., 30* (4): 1328-35.

Rathfelder, K. M., L. M Abriola, T. P. Taylor, and K. D Pennell. 2000. "Surfactant Enhanced Recovery of Tetrachloroethylene from a Porous Medium Containing Low Permeability Lenses 2. Numerical Simulation." submitted to *J. Cont. Hydrol.*

Taylor, T.P. 1999. "Characterization and Surfactant Enhanced Remediation of Organic Contaminants in Saturated Porous Media", Ph.D. dissertation, Georgia Institute of Technology.

Taylor, T.P., Pennell, K.D., Abriola, L.M., and Dane, J.H., 2000a. "Surfactant Enhanced Recovery of Tetrachloroethylene from a Porous Medium Containing Low Permeability Lenses 1. Experimental Studies." submitted to *J. Cont. Hydrol.*

THE INTERPRETATION AND ERROR ANALYSIS OF PITT DATA

Minquan Jin and Richard E. Jackson
Duke Engineering and Services
Austin, TX 78758

Gary A. Pope
University of Texas at Austin
Austin, TX 78712

Abstract: The usual method of partitioning interwell tracer test (PITT) data analysis involves eliminating the scatter in the GC measured tracer concentration data by fitting smooth curves to the data. The NAPL saturation is then estimated based on these smooth curves. Clearly, the curve fitting process will affect the accuracy of the NAPL saturation estimates. The purpose of this paper is to present a PITT data interpretation procedure which uses a general curve fitting equation that can minimize the estimation errors. It demonstrates that the uncertainty in NAPL saturation estimates from a given set of GC measured tracer data can be estimated based on the standard errors of the fitting parameters of the respective fitted curves. An example PITT data analysis is given to illustrate the data interpretation and error analysis process.

INTRODUCTION

PITTs have been used with success in the field to characterize the nonaqueous phase liquid (NAPL) volume and saturation distribution in subsurface (Jin et al., 1997; Annable et al., 1998; Deeds, et al. 1999; Falta et al., 1999; Meinardus et al. 1999; Young, et al., 1999; Deeds, et al. 2000). The PITT consists of the simultaneous injection of several tracers, with different partition coefficients with respect to the NAPL contamination, at one or more injection wells and the subsequent measurement of tracer concentrations at one or more extraction or monitoring wells. Chromatographic separation of the tracers indicates the presence of NAPL in the interwell zone and is used to determine the volume and distribution of NAPL present.

The measurement of tracer concentrations by GC will always include some scatter in the data. This scatter in the tracer concentration data affects the estimation of NAPL saturation. In practice, we fit the measured tracer concentration data of the conservative and partitioning tracers to smooth tracer response curves. The NAPL saturation and retardation factors are then estimated based on these smooth curves. The question is, therefore, what is the uncertainty in the estimated NAPL saturation from a given set of GC measured tracer concentration data.

The purpose of this paper is to: (1) present a flexible curve fitting equation which have been found to fit the majority of tracer data sets with only a small number of fitting parameters and (2) develop a procedure for PITT data analysis that determines the NAPL saturation as well as the uncertainty in the estimate.

METHOD OF TRACER DATA ANALYSIS

The analysis of PITTs is a multi-step process. The analytical tracer concentration data from the laboratory measurements are first incorporated into a sampling database and evaluated through a Quality Assurance/Quality Control (QA/QC) process before being used for the PITT analyses. Tracer concentration data that does not meet QA/QC criteria are excluded from analysis. The QA/QC process also excludes tracer concentration data with measured concentrations below the detection limit of the method of analysis. The remaining data for each tracer test are then evaluated to select a pair of nonpartitioning and partitioning tracers to use to estimate DNAPL volume and saturation in the swept pore volume.

Tracer Data Interpretation Method: The theoretical foundation for the method of first temporal moment analysis of partitioning tracer tests is presented fully in Jin et al. (1995) and Jin (1995). This method is normally used to estimate the swept pore volume, the average DNAPL saturation, and the total DNAPL volume within the tracer swept pore volume. The equations reviewed in this section are for the purpose of providing the reader with a context for the discussion of the PITT error analysis that follows.

The partitioning of a tracer between the aqueous and NAPL phases is described by the equilibrium partition coefficient (K), defined as:

$$K = \frac{C_{tracer,NAPL}}{C_{tracer,water}} \tag{1}$$

where $C_{tracer,NAPL}$ is the tracer concentration in the NAPL, and $C_{tracer,water}$ is the tracer concentration in the aqueous phase. Conservative or nonpartitioning tracers have a partition coefficient of zero relative to the NAPL. As partitioning tracers flow through a NAPL contaminated zone, the partitioning of the tracers between the NAPL and the ground water retards their progress relative to conservative tracers which are unaffected by the presence of NAPL. This retardation is manifested at an extraction well as a separation between the partitioning and nonpartitioning tracer breakthrough curves. The magnitude of this separation is expressed as the retardation factor:

$$R_f = \frac{\bar{t}_p}{\bar{t}_n} \tag{2}$$

where $\bar{t}_p$ and $\bar{t}_n$ are the first temporal moments of the tracer response curves of the partitioning tracer and nonpartitioning tracer, respectively. A temporal moment can be thought of as the centroid of the area under a tracer curve, and for both the partitioning and nonpartitioning case is calculated as:

$$\bar{t}_p = \frac{\int_0^{t_f} tC_p(t)dt}{\int_0^{t_f} C_p(t)dt} - \frac{t_s}{2},$$

(3)

and

$$\bar{t}_n = \frac{\int_0^{t_f} tC_n(t)dt}{\int_0^{t_f} C_n(t)dt} - \frac{t_s}{2}$$

(4)

where t_s is the slug size (the time period over which the tracer mass was injected), t_f is the time at which the tracer test ended, and $C_p(t)$ and $C_n(t)$ represent the partitioning and nonpartitioning tracer concentration as a function of time, respectively. The degree of retardation is a function of the partition coefficient of the tracer and how much NAPL the tracer encountered in the swept pore volume. The average NAPL saturation (S_N) in the tracer swept volume is calculated as:

$$S_N = \frac{R_f - 1}{R_f + K - 1}$$

(5)

To analyze the PITT, the selected tracer data sets are first fit with smooth curves using the following exponential decline equation:

$$C(t) = \sum_{i=1}^{n} \exp(a_i + \frac{b_i}{t} + c_i \ln(t))$$

(6)

where n is an integer, normally with a value of 1, 2 or 3, and a_i, b_i, and c_i are the corresponding fitting parameters.

To estimate the NAPL saturation accurately, the tails of the tracer response curves should be complete. In practice however, the ends of the tracer response curves at the end of the test are often truncated due to dilution of the tracer concentration below the detection limit. However, the tracer response curves can be extrapolated with an exponential function provided the duration of the test is sufficient to establish the decline of the tail. The first moments of truncated tracer response curves can be obtained by dividing the data into two parts. The first part of the tracer curve represents the data from zero to the time t_b, where the response vs. time becomes exponential, and the second covers the exponential part from t_b to infinity. After time t_b, the tracer response has been shown to follow an exponential decline given by:

$$C = C_b e^{-(\frac{t-t_b}{a})}$$

(7)

where 1/a is the slope of a straight line when the tracer response curves are plotted on a semi-log scale, and C_b is the tracer concentration at time t_b, when the curve becomes exponential.

The first moment ($\bar{t}$) (equations 3 and 4) of the partitioning and nonpartitioning tracers can then be calculated as:

$$\bar{t} = \frac{\int_0^{t_b} tCdt + a(a+t_b)C_b}{\int_0^{t_b} Cdt + aC_b} - \frac{t_s}{2}$$

(8)

In summary, the following procedures should be used to estimate the average NAPL saturation from a partitioning interwell tracer test. First, the tracer concentration data for the selected tracer pairs is fitted using equation (6), and , if necessary, extrapolated using equation (7). The first moment of each of the fitted tracer curves is then estimated by either using equations (3) and (4) or equation (8) if the data extrapolation is used. Next, equation (2) is used to calculate the retardation factor. Finally, equation (5) is used to estimate the average NAPL saturation.

Uncertainty in PITT Estimates: An examination of equation (5) indicates that there are two main sources of error associated with the analysis of PITTs that propagate uncertainty to the estimates of average saturation and NAPL volume. The first source of error is an uncertainty in the estimation of the retardation factor of the actual tracer curves. This uncertainty is a function of the accuracy of the analytical method used to measure tracer concentrations. The second source of error is due to the uncertainty in the measurement of the partition coefficient for each tracer.

The uncertainty in the retardation factor is a result of the experimental error in the tracer concentrations. The fitting of a curve (equation 6) to the tracer data provides a way of estimating the uncertainty in the retardation factor by determining the standard error of each of the fitting parameters. This technique is analogous to calculating the correlation coefficient of a linear regression. The accuracy of the retardation factor can be increased by minimizing the standard error of the fitted curve in much the same manner as with a fit to a straight line.

The error associated with the measurement of the partition coefficient can be estimated based on the error analyses of numerous laboratory partition coefficient measurements. Based on our own laboratory experimental results, it was found that the average relative error in the partition coefficient measurement is about 10%. In practice, the errors associated the partition coefficient measurements should be estimated for each set of laboratory data as part of the experimental investigation using representative NAPL samples from the site of interest.

The uncertainty in the PITT analysis therefore can be calculated from the standard error incorporated in both the retardation factor estimate, and the uncertainty in the partition coefficient. Assuming that the standard errors of the first moments of the nonpartitioning and partitioning tracer response curves are $\Delta \bar{t}_p$ and $\Delta \bar{t}_n$ respectively, the standard error of the retardation factor can be estimated based on the theory of error propagation. From equation (2), we have,

$$\Delta R_f \leq \sqrt{\left(\frac{\partial R_f}{\partial \bar{t}_p}\Delta \bar{t}_p\right)^2 + \left(\frac{\partial R_f}{\partial \bar{t}_n}\Delta \bar{t}_n\right)^2}$$

$$= \sqrt{\left(\frac{\Delta \bar{t}_p}{\bar{t}_n}\right)^2 + \left(\frac{R_f \Delta \bar{t}_n}{\bar{t}_n}\right)^2}$$

(9)

Similarly, we have from equation (5),

$$\Delta S_D \leq \sqrt{(\frac{\partial S_D}{\partial R_f}\Delta R_f)^2 + (\frac{\partial S_D}{\partial K}\Delta K)^2}$$

$$= \sqrt{(\frac{K}{(R_f + K - 1)^2}\Delta R_f)^2 + (\frac{R_f - 1}{(R_f + K - 1)^2}\Delta K)^2}$$

(10)

Once the standard error of the retardation factor is calculated using equation (9), the uncertainty in the DNAPL saturation estimation can be estimated using equation (10).

EXAMPLE APPLICATION

To illustrate the procedure for data analysis described above, an example PITT data analysis follows. The data is from an extraction well used during a full scale PITT recently completed at Operable Unit 2 (OU2), Hill AFB in Utah. Detailed information on the implementation and analysis of the full-scale field PITT from which the data is taken can be found in Meinardus et al. (1999) and USAFB (1999). The PITT characterizes an entire DNAPL source zone involving a total swept pore volume of over one million liters (~285,000 gallons).

The tracer response data from extraction well EW5 is shown in Figure 1. In this figure, the symbols represent the tracer concentrations as measured on a GC. The smooth curves are generated using equation (6). The conservative tracer 1-propanol and the partitioning tracer n-heptanol (K=85.6 with the Hill OU2 DNAPL) were chosen as the tracers to be used to estimate the DNAPL saturation in the swept pore volume. The choice of these tracers was based on a QA/QC data analysis as outlined above. The fitted equations for 1-propanol and n-heptanol are,

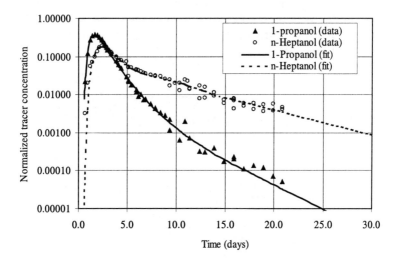

FIGURE 1. Normalized tracer concentration data and corresponding fitted curves for extraction well EW5

$$C_1(t)_{propanol} = Exp(14.24 - \frac{8.99}{t} - 5.63\ln(t))$$

and

$$C_1(t)_{hep\,tan\,ol} = Exp(15.96 - \frac{14.25}{t} - 5.93\ln(t)) + Exp(25.01 - \frac{59.41}{t} - 7.18\ln(t))$$

For this example, the tracer concentration data was extrapolated using the exponential decline equation (7) to extrapolate the tracer data beyond day 15. The generalized extrapolation equations for 1-propanol and n-heptanol, respectively, are,

$$C_2(t)_{propanol} = 0.2 * Exp(-\frac{t-15}{3.4})$$

and

$$C_2(t)_{propanol} = 5.8 * Exp(-\frac{t-15}{6.7})$$

Based on the above curve fitting and data extrapolation equations, the first temporal moments of the response cureves for 1-propanl and n-heptanol were estimated using equation (8). The standard errors in the first moment estimation of the tracer curves were then estimated based on the difference in the fitted values and the actual values obtained from the field. The results are summarized in Table 1. Once the first temporal moments of the 1-propanl and n-heptanol response curves were estimated, the retardation factor and the DNAPL saturation were obtained using equations (2) and (5), respectively. The uncertainties associated with these estimated were then estimated using equations (9) and (10). The results are tabulated in Table 1.

Table 1. Summary of tracer data interpretation and error analysis results

$\bar{t}_n + \Delta \bar{t}_n$ (days)	$\bar{t}_p + \Delta \bar{t}_p$ (days)	$R_f \pm \Delta R_f$	$\Delta R_f / R_f$ (%)	$S_D \pm \Delta\, S_D$ (%)	$\Delta S_D / S_D$ (%)
2.08 ± 0.03	5.64 ± 0.04	2.7 ± 0.043	1.6	2.0 ± 0.2	10.0

As the results in Table 1 indicate, the uncertainty related to the retardation factor estimate for this particular example is ±0.043, which translates into a relative error of 1.6%. The estimated uncertainty of average NAPL saturation estimation is ±0.2, which translates into a relative error of 10%.

CONCLUSION AND DISCUSSION

In this paper, we have presented a PITT data interpretation analysis procedure using a flexible curve fitting equation. It demonstrates that the uncertainty in the estimate of NAPL saturation using a given set of tracer concentration data can be calculated. An example PITT data analysis is given to illustrate the data interpretation and error analysis process.

ACKNOWLEDGMENTS

The example application presented in this paper is based on the results from the DNAPL Source Delineation Project funded by the Environmental Restoration Division (EMR) of the Hill Air Force Base (AFB). The authors would like to thank each individual involved for their support for this project.

REFERENCES

Annable, M., Rao, P.S.C., Hatfield, K., Graham, W.D., and Enfield, C.G. 1998. "Partitioning tracers for measuring residual NAPL: Field-Scale Test Results." *J. Environmental Engineering* 124(6), 498-503.

Deeds, N.E.; McKinney, D.C.; and Pope, G.A. 1999. "Vadose Zone Characterization at a Contaminated Field Site Using Partitioning Interwell Tracer Technology." *Environmental Science & Technology*, 33(16): 2745-2751

Deeds, N.E., McKinney, D.C., and Pope, G.A., 2000. "Laboratory Characterization of Non-aqueous Phase Liquid/Tracer Interaction in Support of a Vadose Zone Partitioning Interwell Tracer Test." *Journal of Contaminant Hydrology*, 41: 193-204.

Falta, R. W.; Lee, C. M.; Brame, S. E.; Roeder, E.; Coates, J. T.; Wright, C.; Wood, A. L.; and Enfield, C. G., 1999. "Field Test of High Molecular Weight Alcohol Flushing for Subsurface Nonaqueous Phase Liquid Remediation." *Water Resources Research*. 35(7), 2095-2108.

Jin, M.; Delshad, M.; McKinney, D.C.; Pope, G.A.; Sepehrnoori, K.; Tilburg, C.E.; Jackson, R.E., 1995. "Partitioning Tracer Test for Detection, Estimation and Remediation Performance Assessment of Subsurface Nonaqueous Phase Liquids." *Water Resources Research*. 31(5), 1201-1211.

Jin, M, Jackson, R.E., Pope, G.A., Taffinder, S., 1997, "Development of Partitioning Tracer Tests for Characterization of Nonaqueous-Phase Liquid-Contaminated Aquifers." In *Proceeds of SPE 72nd Annual Technical Conference & Exhibition*, San Antonio, Texas, Oct. 1997.

Jin, M., 1995. *A Study of Nonaqueous Phase Liquid Characterization and Surfactant Remediation*. Ph.D. Dissertation, U. of Texas, Austin.

Meinardus, H. W., Fort, M., Jackson, R. E., Jin, M. Ginn, J. S., Stotler, G.C., 1999. "Delineation of DNAPL Source Zone with Partitioning Interwell Tracer Tests." in *Proceedings of the 2000 Petroleum Hydrocarbons and Organic Chemicals in Ground Water: Prevention, Detection, and Remediation*, Houston, TX, Nov., 1999, 79-85.

Young, C.M., Jackson, R. E., Jin, M. Londergan, T. J., Mariner, P. E., Pope, G. A., Aderson, F. F., and Houk, T., 1999. "Characterization of a TCE DNAPL zone in Alluvium by Partitioning Tracers." *Ground Water Monitoring and Remediation*. 21(1), 84-94.

United States Air Force (USAF), 1999. *Dense Nonaqueous Phase Liquid (DNAPL) Source Delineation Project Final Report, Operable Unit 2*

MULTIPHASE FLOW SIMULATION IN PHYSICALLY AND CHEMICALLY HETEROGENEOUS MEDIA

Heejun Suk (Penn State University, University Park, PA)
G. T. Yeh (Penn State University, University Park, PA)

ABSTRACT: The multiphase flow simulator, MPS, is developed based on the fractional flow approach considering the fully three phase flow in physically and chemically heterogeneous media. Most existing models are limited to two phase flow when considering the physically heterogeneous media. Also, these models have mostly focused on the water-wet media. A number of factors including variations in aqueous chemistry and mineralogy might affect the wettability change from water-wet to oil-wet media. Furthermore, the wetting of porous media by oil could be heterogeneous, or fractional, rather than uniform due to the heterogeneous nature of subsurface media and the factors that influence wettability. However, in this study, chemically heterogeneous media considering fractional wettability is simulated with MPS.

INTRODUCTION

Historically, there have been two main approaches to modeling multiphase flow, arising in the disciplines of hydrology and petroleum engineering. The first is based on individual balance equations for each of the fluids, while the second involves manipulation and combination of those balance equations into modified forms, with concomitant introduction of ancillary functions such as the fractional flow function. The first approach is referred to as pressure-based, while the second approach is referred to as fractional flow-based. The pressure approach has been widely used in the hydrologic literature. In this approach, the governing equations are written in terms of the pressures through a straightforward substitution of Darcy's equation into the mass balance equations for each phase. This approach has been adopted by a number of authors, (Pinder and Abriola, 1986; Sleep and Sykes, 1989; Kaluarachchi and Parker, 1989; Celia and Binning, 1992; Kueper and Frind, 1991(a) and (b)). The fractional flow approach originated in the petroleum engineering literature employs saturation of phases and a pressure as the independent variables. This approach leads to three equations: one pressure and two saturation equations. The fractional approach has been employed by several authors (Guarnaccia and Pinder, 1997; Binning and Celia, 1999). However, until now most fractional flow approaches have been limited to two-phases and specific boundary conditions in only physically heterogeneous media. Usually, boundary conditions are the Dirichlet-type, which is not practical for real problems because it is impossible to know the pressure of infiltrating phases over time.

In this paper, a two-dimensional finite element model to study the simultaneous movement of nonaqueous phase liquid, water, and gas in physically and chemically heterogeneous porous media is developed. Pressure equation is solved by using the standard Galerkin finite element method. The upstream finite element method is employed to solve two saturation equations. In order to obtain simultaneous solution of saturation equations, we use the BiCG-STAB method with incomplete modified cholesky decomposition as a preconditioner. A distinctive feature is that the phase flow equations are given in a fractional flow formulation, i.e., in terms of saturation and a total pressure. These variables exist throughout the solution domain regardless of whether the nonwetting phase is present. This eliminates the need to specify a small, fictitious degree of nonwetting saturation where only the wetting phase is present, as is normally required with more traditional pressure-based simulators.

GOVERNING EQUATIONS

The fractional flow approach is based on the following equations. The pressure equation is as follows

$$-\nabla \bullet \kappa \left(P_t + \bar{\rho} g \nabla z \right) = 0 \tag{1}$$

where P_t is the total pressure, $\bar{\rho}$ is the mobility weighted average density, and total mobility κ is defined by

$$\kappa = \mathbf{k} \left(\frac{k_{r1}}{\mu_1} + \frac{k_{r2}}{\mu_2} + \frac{k_{r3}}{\mu_3} \right) \tag{2}$$

in which $\mathbf{k}$ is the intrinsic permeability; k_{r1}, k_{r2}, and k_{r3} are the relative permeability of water, NAPL, and air phase; μ_1, μ_2, and μ_3 are the viscosity of water, NAPL, and air phase. Total flux, $\vec{V}_t$, defined as the sum of the phase velocities is given by

$$\vec{V}_t = -\kappa \left[\nabla P_t + \bar{\rho} g \nabla z \right] \tag{3}$$

Two saturation equations are given by

$$\frac{\partial n_d S_1}{\partial t} + \vec{V}_t \bullet \frac{d\kappa_1}{dS_1} \nabla S_1 = -\kappa_1 \nabla \bullet V_t +$$

$$\nabla \bullet \kappa_1 \kappa \left[\left(\kappa_2 \frac{\partial P_{C12}}{\partial S_1} + \kappa_3 \frac{\partial P_{C13}}{\partial S_1} \right) \nabla S_1 + \left(\kappa_2 \frac{\partial P_{C12}}{\partial S_t} + \kappa_3 \frac{\partial P_{C13}}{\partial S_t} \right) \nabla S_t + \left(\rho_1 - \bar{\rho} \right) g \nabla z \right] \tag{4}$$

$$-\frac{\partial n_d S_t}{\partial t} + \vec{V}_t \bullet \frac{d\kappa_3}{dS_t} \nabla S_t = -\kappa_3 \nabla \bullet V_t +$$

$$\nabla \bullet \kappa_3 \kappa \left[\left(\kappa_2 \frac{\partial P_{C32}}{\partial S_1} + \kappa_1 \frac{\partial P_{C31}}{\partial S_1} \right) \nabla S_1 + \left(\kappa_2 \frac{\partial P_{C32}}{\partial S_t} + \kappa_1 \frac{\partial P_{C31}}{\partial S_t} \right) \nabla S_t + \left(\rho_3 - \bar{\rho} \right) g \nabla z \right] \tag{5}$$

where n_d is the diffusive porosity; S_1, S_2, and S_3 are the water, NAPL, and gas phase saturation; P_{C12} is the capillary pressure of water and NAPL; P_{C13} is the capillary pressure of water and gas; P_{C32} is the capillary pressure of gas and NAPL; P_{C31} is the capillary pressure of gas and water; κ_1, κ_2, and κ_3 are mobilities of water, NAPL, and gas phases; and $S_t = S_1 + S_2$ is total liquid saturation.

Equations (1), (4), and (5) must be supplemented with appropriate initial boundary conditions. Initial boundary conditions for each phase are written as

$$S_i = S_i(x,z) \text{ on R for i=1,3} \tag{6}$$

Boundary conditions can be written as combination of two types of boundaries of individual phases, flux-type and Dirichlet type boundaries.

$$P_i = P_{ni}(x_b,z_b,t) \text{ on } B_{di} \tag{7}$$
$$V_i \bullet \vec{n} = q_{ni}(x_b,z_b,t) \text{ on } B_{fi} \tag{8}$$

where (x_b,z_b) is the spatial coordinate on the boundary; $\vec{n}$ is outward unit vector normal to the boundary; P_{ni} and q_{ni} are the prescribed Dirichlet functional value and flux value of phase i, respectively; B_{di} and B_{fi} are the Dirichlet and flux boundaries for the i-th phase, respectively.

VERIFICATION

In general, analytical solutions incorporating fully the effects of gravity and capillarity in transient multiphase flow through heterogenous porous media are not tractable. Many previous authors (Binning and Celia, 1999; Yeh et al., 1996) have chosen to verify their models against the classical Burkley-Leverett problem (Burkley and Leverett, 1942) which represents one dimensional, horizontal two-phase flow in the absence of capillary forces. In this study, model verification is carried out by comparison to an analytical solution, based on one by McWhorter and Sunada (1990), which fully incorporates the effect of capillarity. The physical scenario used here for model verification involves a one-dimensional, 2000 cm long horizontal column of a porous media initially completely saturated by an incompressible wetting fluid, water. A nonwetting fluid is continuously injected at the inflow end of column for t > 0 at a rate such that a constant saturation ($S_1(0,t) = 0.84$) is maintained at this boundary. To effect a constant saturation at this boundary, flux of nonwetting phase is prescribed inversely proportional to square root of time. The boundary and initial conditions for the displacement of a wetting phase by entry of a nonwetting phase are

$$q_{nw}(0,t) = At^{-1/2} \quad t > 0 \tag{9}$$

$$S_1 (\infty, t) = S_i \quad t > 0 \tag{10}$$
$$S_1 (x, 0) = S_i \tag{11}$$

where q_{nw} is the flux of nonwetting phase, the parameter A is a constant. In this study, $S_i = 1$ and $A = 2 \ cm / day^{\frac{1}{2}}$ are chosen. For these conditions the analytical solution is valid up to about 249 days, at which point the nonwetting front first reaches the end of the column. The input parameter used in this verification is indicated in Table 1. Figure 1 illustrates the excellent agreement between the numerical model and the analytical solution for the distribution of fluid saturations in the column at various times.

TABLE 1. Input parameters for model verification.

k (cm²)	n_d	μ_1(g/cm/day)	μ_2(g/cm/day)	μ_3(g/cm/day)	α_{21}(cm⁻¹)	α_{32}(cm⁻¹)
5.0×10^{-7}	0.35	864	432	15.812	0.0792	0.0924

α_{31}(cm⁻¹)	n (Van Genuchten fitting paramter)	Δx (cm)	Initial time step size (day)	Increment of time step	Maximum time step size (day)	Horizon. Distance (cm)
0.044	5.0	20	1.0×10^{-5}	0.1	1.0	2000

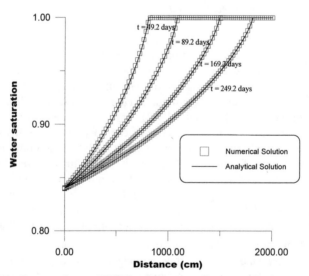

FIGURE 1. Comparison of MPS solutions and exact solutions at various times.

RESULTS AND DISCUSSION

The infiltration of a DNAPL in a fully water-saturated porous media is simulated to assess the effect of physically heterogeneous and chemically heterogeneous media.

DNAPL Flow in Physically Heterogeneous Media.

It is assumed that a low permeable lens is placed in the interior of the simulation region to investigate capillary pressure effects on DNAPL flow. Capillary pressure relations after van Genuchten model (Parker et al., 1987) with different pore size distribution index are assumed in the lens and surrounding matrix. The bottom of the reservoir is impermeable for all three phases. Hydrostatic pressure distribution for water pressure and atmospheric pressure for air phase are prescribed at the left and right boundaries, while for NAPL phase, pressure which makes NAPL saturation equals to zero is prescribed through using van Genuchten model (Parker et al., 1987). In other word, pressure for NAPL phase is equal to air pressure. At the inlet on the top boundary, a flux, 10.0 cm/day for the nonwetting phase is prescribed while for water and air phase no flow condition is prescribed. Initial conditions are: hydrostatic pressure distribution for water, atmospheric condition for air phase, and pressure for NAPL to make $S_2 = 0$. The fluid parameters are $\rho_1 = 1.00$ g/cm^3, $\rho_2 = 1.63$ g/cm^3, $\rho_3 = 1.2 \times 10^{-3}$ g/cm^3, $\mu_1 = 864.0$ g/cm/day, $\mu_2 = 860.6$ g/cm/day, and $\mu_3 = 15.228$ g/cm/ day. Relative permeabilities and capillary pressure are defined after the model of van Genuchten (Parker et al., 1987). Two different van Genuchten parameter n for lens is assumed to investigate the effect of the capillary barrier on DNAPL flow: Case A and Case B. Soil properties for case A and B are shown in Table 2. Simulation results for case A with high n is depicted in Figure 2(a), and for case B with low n in Figure 2(b). Low value of n indicates wide pore size distribution which causes lower entry pressure of nonwetting phase. It means that infiltration in the lens is possible for the lower entry pressure. As shown in Figure 2(a), it is observed that no DNAPL infiltrates the fine sand lens while NAPL saturation is discontinuous over the lens boundary. Whereas, in case B, DNAPL infiltration is possible to make NAPL saturation continuous through the lens.

TABLE 2. Soil properties for the infiltration problem.

	k (cm^2)	n_d	α_{21} (cm^{-1})	α_{32} (cm^{-1})	α_{31} (cm^{-1})	n (van Genuchten Fitting parameter)
Matrix (Case A)	6.64×10^{-6}	0.43	0.009	0.0105	0.005	2.80
Lens (Case A)	3.32×10^{-8}	0.36	0.009	0.0105	0.005	2.80
Matrix (Case B)	6.64×10^{-6}	0.43	0.009	0.0105	0.005	2.80
Lens (Case B)	3.32×10^{-8}	0.36	0.009	0.0105	0.005	1.80

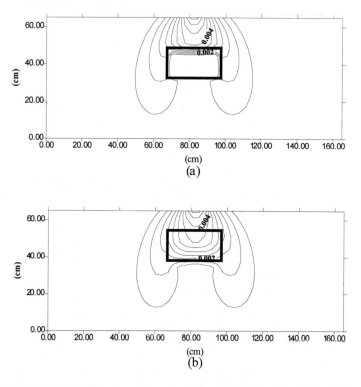

FIGURE 2. DNAPL saturation plots for the infiltration after 630 steps of
1.0×10^{-4} **day for cases A and B.**

DNAPL Flow in Chemically Heterogeneous Media.
    In this section DNAPL infiltration into a physically homogeneous media
having various chemical characterisitcs will be presented. Here, we assume that
chemical heterogeneties are indicated by contact angle which is an index of
wettability, and is defined as the angle between the solid-water and water-organic
contact lines. When $\phi_{contact} < 90$, water is classified as the wetting fluid, whereas,
for $\phi_{contact} > 90$ the water is the nonwetting fluid. Relationship between contact
angle and capillary pressure have been defined by Bradford et al. (1998). This
relationship is adopted in this study. Numerical experiments are performed for
four different contact angles, $\phi_{contact}$ = 0°, 60°, 120°, and 180°. Simulation
conditions are same as Figure 2(a) except homogeneous soil property and soil
surface wetting characteristics. As shown in Figure 3, depth of DNAPL
infiltration is lower and organic saturation are higher as $\phi_{contact}$ increases. This can
be explained that as $\phi_{contact}$ increases, capillary pressure and relative permeability
of NAPL is decreased, hence the mobility of the NAPL decreases. Therefore
porous medium retard the NAPL more effectively.

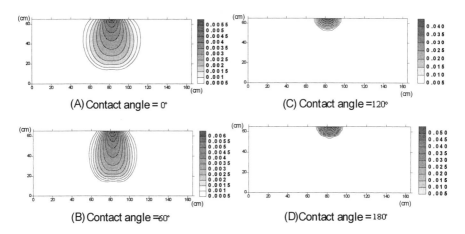

FIGURE 3. DNAPL saturation at simulation time 2.58×10^{-2} days at four different contact angles in the chemically heterogeneous media.

CONCLUSIONS

Two-dimensional simulator, MPS based on fractional flow approach is developed considering the fully three phase flow in physically and chemically heterogeneous media. The developed program is verified with an exact analytical solution which fully incorporates effects of capillarity and relative permeability. In order to investigate the effects of physically heterogeneous and chemically heterogeneous media on DNAPL flow, numerical experiments are performed. Chemically heterogeneous media is characterized by relationship between contact angle and capillary pressure defined by Bradford et al. (1998). Simulation results in Figure 2 show that in physically heterogeneous media having wide pore size distribution DNAPL infiltration into low permeability lens is possible. The reason is that wide pore size distribution causes nonwetting entry pore pressure relatively lower than narrow pore size distribution. The lower entry pore pressure can make DNAPL infiltrate into low permeability lens. In the chemically heterogeneous media, as shown by Figure 3, as $\phi_{contact}$ increases, depth of DNAPL infiltration is lower and organic saturation are higher. This can be explained that increasing $\phi_{contact}$ makes capillary pressure and relative permeability decrease, hence DNAPL flow is retained by porous media greatly. In summary, we can check that capillary barrier effects on organic liquid migration are known to occur at varying soil surface wetting characteristics as well as at soil textures, due to abrupt changes in pore size distributions. This fact is confirmed with the previous studies.

REFERENCES

Binning, P., and M. A. Celia. 1999. "Practical Implementation of the Fractional Flow Approach to Multi-Phase Flow Simulation." *Advances in Water Resources.* 22: 461-478.

Bradford, S. A., L. M. Abriola, and K.M. Rathfelder. 1998. "Flow and entrapment of dense nonaqueous phase liquids in physically and chemically heterogeneous aquifer formations." *Advances in Water Resources.* 22(2): 117-132.

Burkley, S. E., and M. C. Leverett. 1942. "Mechanism of Fluid Displacement in Sands." *Trans. AIME.* 146: 107-116.

Celia, M. A., and P. Binning. 1992. "Two-Phase Unsaturated Flow: One Dimensional Simulation and Air Phase Velocities." *Water Resour. Res..* 28: 2819-2828.

Guarnaccia, J. F., and G. F. Pinder. 1997. NAPL: A Mathematical Model for the Study of NAPL Contamination in Granular Soils, Equation Development and Simulator Documentation. The University of Vermont, RCGRD #95-22.

Kaluarachchi, J. J., and J. C. Parker. 1989. "An Efficient Finite Elements Method for Modeling Multiphase Flow." *Water Resour. Res..* 25: 43-54.

Kueper, B. H., and E. O. Frind. 1991a. "Two-Phase Flow in Heterogeneous Porous Media, 1. Model Development." *Water Resour. Res..* 27: 1049-1057.

Kueper, B. H., and E. O. Frind. 1991b. "Two-Phase Flow in Heterogeneous Porous Media, 2. Model Application." *Water Resour. Res..* 27: 1058-1070.

McWhorter, D. B., and D. K. Sunada. 1990. "Exact Integral Solutions for Two Phase Flow." *Water Resour. Res..* 26(3): 399-414.

Parker, J. C., R. J., Lenhard, and T. Kuppusamy. 1987. "A Parameteric Model for Constitutive Properties Governing Multiphase Flow in Porous Media." *Water Resour. Res..* 23: 618-624.

Pinder, G. F., and L. M. Abriola. 1986. "On the Simulation of Nonaqueous Phase Organic Compounds in the Subsurface." *Water Resour. Res..* 22: 109-119.

Sleep, B. E., and J. F. Sykes. 1989. "Modeling the Transport of Volatile Organics in Variably Saturated Media." *Water Resour. Res..* 25(1): 81-92.

Yeh, G.T., H. P., Cheng, J. R. Cheng, T. E. Short, and C. Enfield. 1996. An Adaptive Local Grid Refinement Algorithm to Solve Nonlinear Transport Problems with Moving Fronts. Proc. XI-th Int. Conf. On Numerical Methods in Water Resources, Cancun, Mexico, July 22-26, p. 577-584.

SIMULATION OF DNAPL DISTRIBUTION RESULTING FROM MULTIPLE SOURCES

Michael Fishman (Dynamac Corp., Ada, OK); Joseph Guarnaccia (Ciba Specialty Chemicals Corp., Toms River, NJ); Lynn Wood (US EPA/NRMRL, Ada, OK); Carl Enfield (US EPA/NRMRL, Cincinnati, OH)

ABSTRACT: A three-dimensional and three-phase (water, NAPL and gas) numerical simulator, called NAPL, was employed to study the interaction between DNAPL (PCE) plumes in a variably saturated porous media. Several model verification tests have been performed, including a series of 2-D laboratory experiments involving the migration of PCE through a variably saturated, homogeneous sand. A comparison of the experimental data to the model results illustrates the effect and importance of fluid entrapment and saturation hysteresis.

The NAPL model was used to simulate a 3-D multi point PCE source release within a contained test cell at the Groundwater Remediation Field Laboratory (GRFL) in Dover, Delaware. In this experiment, the migration of PCE in the unsaturated and saturated zones, under various infiltration scenarios, was simulated. The modeling of multiple injection points in a homogeneous aquifer shows that the ultimate distribution of PCE depends on the injection point locations and the time-varying release rates, and the depth to the water table. In general, an intermittent, slow, injection rate caused narrow, deeply penetrating DNAPL plumes. On the other hand, higher injection rates resulted in a wider horizontal distribution and more interaction between neighboring plumes, thus creating non-symmetric distributions and an increase in the flow rate and depth of penetration.

INTRODUCTION

A 3-D NAPL model was developed to investigate the movement of organic compounds in both homogeneous and heterogeneous porous media. Particular attention was paid to the development of a sub-model that describes three-phase hysteretic permeability-saturation-pressure relationships, and the potential entrapment of fluids when they are displaced. Several laboratory experimental data sets have been compared to simulator predictions, including a series of 2-D laboratory experiments involving PCE release conducted by the authors at the R. S. Kerr Laboratory, Ada OK, and described in Fishman et al. (1998) and Guarnaccia et al. (1997).

The results of modeling runs were used to develop a DNAPL release strategy for remediation technology demonstrations at the Dover National Test Site (DNTS), Dover, Delaver, and to predict the DNAPL movement from injection points under a range of hydrodynamic release conditions. Because the hydrodynamic properties and spatial variability of these properties strongly influences the behavior of DNAPL in the subsurface, simulations were done for a range of hydrodynamic conditions in both homogeneous and heterogeneous porous media.

The attributes (parameters, processes) that control DNAPL distribution in subsurface granular porous media include: fluid properties (density, viscosity, interfacial tension, wettability), soil properties (hydraulic conductivity, heterogeneity), source conditions (release rate and proximity of multiple sources). The focus of this paper is on the issue of source conditions: how release rates and source proximity affect DNAPL distribution in the subsurface.

NUMERICAL MODEL

A numerical model, called NAPL (Guarnaccia et al., 1997), was used to simulate the experiment described above. The model has the following conceptual and computational attributes, which are assumed to be relevant to the physical experiment to be modeled:
1) Simultaneous flow of water, NAPL, and gas;
2) The three-phase relative permeability-saturation-capillary pressure (k-S-P) relationships are based on fluid phase wettablility considerations and two-phase data; wettability follows, from most to least, water-NAPL-gas;
3) The three-phase k-S-P model reduces to the appropriate two-phase model when appropriate;
4) The k-S-P model includes flow-path-history-dependent functionals (hysteresis);
5) The k-S-P model includes a mechanism for fluid entrapment during drainage (residual emplacement), where the amount entrapped is a function of the maximum imbibed saturation;
6) The S-P model employs capillary pressure scaling to account for variable porous medium and fluid properties;
7) At each boundary node, one can specify either a no flow condition (can also be coupled with a point source or sink 'well' rate for one or more phases), or any one of the three phase pressures known, or all the primary variables known (i.e., one pressure and two saturations);
8) A numerical 'peclet criterion' can be employed to ensure that the scale of the saturation–capillary pressure functional is compatible to the scale of the grid.

The model employs the collocation finite element method to approximate the system of governing equations spatially, and an implicit finite difference approximation in time. The non-linear system of governing equations is solved using a sequential solution algorithm. Details of the numerical methods can be found in Guarnaccia et al. (1997).

NUMERICAL EXPERIMENTS

Model Testing. A two-dimensional artificial aquifer experiment was conducted to study how DNAPL (PCE) migrates through a variably saturated homogeneous, isotropic, porous medium. A video image of the experiment was analyzed to define the DNAPL saturation at the pixel-scale as a function of time. Once constructed, the image was compared to the solution of the numerical model designed to simulate the same experiment. When the image was averaged to the same spatial scale as the model grid, a qualitatively similar solution was observed.

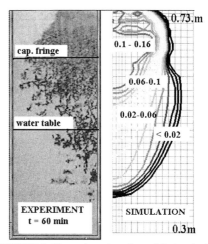

FIGURE 1. Experiment and model simulation.

Figure 1 compares the image of experimental distribution of PCE and the numerical results at T = 60min (as the PCE is migrating through the saturated zone). Note that the solutions are qualitatively similar, indicating that the numerical model effectively captures the physics of the problem as defined at the scale of a grid cell. This information can be used in practice to define appropriate dissolution and vaporization mass transfer rates at the grid scale. Details of the experiment and the video imaging analysis can be found in Fishman et al. (1998).

Effects of Heterogeneity on DNAPL Distribution. It is well known that DNAPL distribution in the subsurface is very sensitive to heterogeneity in hydraulic conductivity (see for example Poulsen and Kueper, 1992, and Kueper et al., 1993). The effect of soil heterogeneity on PCE distribution is illustrated in Figure 2. The same PCE flood experiment was run in a variably saturated domain with three hydraulic conductivity (K) realizations: a homogeneous domain (part a), a horizontal low K lens located below the water table (part b), a vertical high K lens located below the water table (part c). The

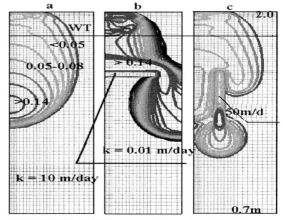

FIGURE 2. PCE saturation distribution. a–homogeneous media. b–single horizontal lense. c– single vertical lense.

low K horizontal lens exerts a very strong influence on the PCE migration pattern. The inability to penetrate this lens is due to a combination of preferential flow through high K sand and the fact that the capillary pressure above the lens is less than the entry pressure for the lens (Figure 2 b). Conversely, the high K vertical lens enhances vertical PCE migration via the combination of preferential flow and the low entry pressure associated with the lens (Figure 2 c).

The simulations show that heterogeneity causes the shape and internal structure of the DNAPL plume to differ significantly from that predicted using the homogeneous realization. At the field scale, because of uncertainties in defining soil properties, and computational issues regarding discretization, DNAPL modeling becomes one of a sequential screening exercise. Specifically, initial modeling with a mean homogeneous K-field is combined with data generated from a groundwater quality profile located down-gradient of the source (see Pardieck and Guarnaccia, 1999). An appropriate 'modeling K-field' is defined using the groundwater quality profile signature.

Effects of Source Rate, and Multiple Release Locations. In this section we present results of several 3-D PCE flood numerical experiments aimed at providing insight into the importance of source area characterization in assessing DNAPL distribution in the subsurface. Specifically, we consider the variables of source release rate and the number and proximity of sources. The 3-D domain is homogeneous, while the source distribution is heterogeneous.

Soil-and Fluid-Phase Parameters. The model parameters, which are presented in Table 1, were determined in the laboratory or obtained from literature.

TABLE 1. Parameters used in the DNAPL spill simulations.

Soil Properties
porosity - 0.37; hydraulic conductivity - 10 m/day; volume density - 1.7 g/cm^3.
van Genuchten k-S-P parameters: a_d - 4.00/m; a_i - 6.00/m; n-6.35;
DNAPL residual saturation as a nonwetting phase - 0.14.
DNAPL properties
viscosity - 0.90 Cp; density - 1.626 g/ml; interfacial tension, PCE -water- 47.5 dyne/cm; interfacial tension, PCE- air - 31.74 dyne/cm.
Water properties
viscosity - 0.99 Cp; density - 0.99 g/ml; interfacial tension, air-water - 72.75 dyne/cm.
Air properties
viscosity - 0.02 Cp; density - 0.00129 g/ml.

The Flow Domain. The dimensions of the flow domain selected for 3-D multi-well simulations are 200 cm high, 460 cm long, and 300 cm wide. The domain is discretized into an irregular grid of 10 by 26 by 24 elements with spacing that varies from 5.0 by 5.0 by 5.0 cm in proximity to the potential source areas to 30.0 by 30.0 by 30.0 cm at the perimeter of the domain. A plan view of the domain is shown in Figure 3 and a cross section is shown in Figure 4.

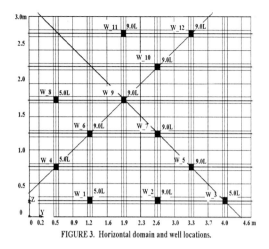

FIGURE 3. Horizontal domain and well locations.

Initial, Point Source, and Boundary Conditions. The domain is initially free of PCE. PCE flood simulations were performed in which PCE was allowed to infiltrate from as many as twelve point sources located below ground surface, but above the water table, at a rate between 0.25 L/min and 0.9 L/min (total volume injected at any one well ranges from 5.0 to 9.0 L). The locations of the point sources are shown in Figure 3. The boundary conditions are no-flow along all boundaries except the top, where a constant atmospheric pressure is prescribed.

The full experiment was modeled as a sequential series of three sub-models. Each sub-model solves a part of the overall flow problem:
1. The initial conditions for sub-model 1 were full water saturation. At time>0, the water table was lowered between 20 and 40 cm and the two-phase system (water and gas) was allowed to approach steady-state conditions. The final distribution of water saturation was adopted as the initial conditions for second sub-model.
2. Using the initial conditions (distribution of water saturation) determined from sub-model 1, DNAPL was released at a predefined rate from the source(s) until a predetermined total volume was released.
3. After the DNAPL was released to the formation the spill was allowed to redistribute for the duration of the five day simulation period.

For multiple simulations involving changing parameters (e.g., source conditions), this structure allows for restart at known intermediate flow conditions, and thus, it is a time saving measure.

RESULTS AND DISCUSSION

Effect of Injection Rate on DNAPL Distribution. Figure 4 (cross-section) shows the solution after a ten minute release of PCE and five days of redistribution from two point sources: at the source on the left, 5.0L of PCE were injected, while at the source on the right 9.0 L were injected. The results show that as the release rate increases the DNAPL plume tends to spread more laterally.

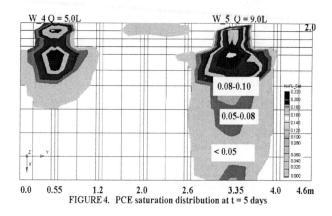

FIGURE 4. PCE saturation distribution at t = 5 days

The higher volume injected on the right and resulted in deeper penetration into the saturated zone. For the case of equivalent volume injected at different rates (not shown), the slower rate resulted in a plume which had a smaller effective radius and greater penetration. This result is consistent with that described in Poulsen and Kueper (1992).

The Multi Wells Simulations. A 3-D simulation was performed in which PCE was allowed to infiltrate from twelve point sources located in the vadose zone (see Figure 3, plan view). Each well injected between 5.0 and 9.0 L of PCE in approximately 10 minutes (the volume for each well is shown in Figure 3). After injection the PCE was allowed to redistribute for five days simulation time.

Below the water table the extent of the PCE plume reached the bottom and in three areas of PCE release (W_2; W_11 and W_12) it accumulated as a 'pool' of free phase (mobility > 1) (Figures 5 and 6). Pools were created due to theproximity of the wells (20cm) to the impervious walls. In the rest of the area, with a release amount of 9.0 L, the plume may reach the bottom but no pool will be created.

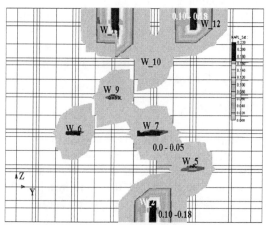

FIGURE 5. Distribution of PCE saturation at depth 200 cm.

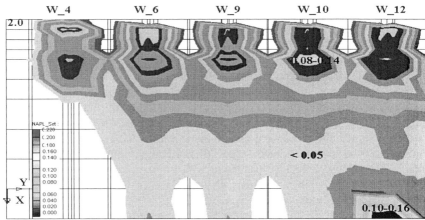

FIGURE 6. Distribution of PCE saturation at cross section W_4 - W_12

Figure 7(a) presents the simulation when distance between wells is 1.0m and less than spreading area (1.4m) shown in part b of Figure 7. In this case we have more interaction between neighboring plumes, thus creating non-symmetric

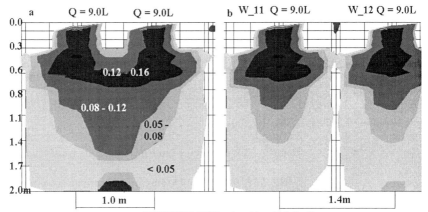

FIGURE 7. PCE saturation distribution.
a - distance between wells is 1.0m. b - distance between wells is 1.4m.

distributions and an increase in the flow rate and depth of penetration. The free phase of DNAPL will accumulate at the bottom of the cell.

CONCLUSIONS

Three important conclusions have been drawn from the modeling experiments:
   a) The depth and the width of spreading PCE are influenced by the release rate and the amount of release. A smaller rate of PCE release gives deeper

penetration and a smaller radius of influence than from an instantaneous spill. Our laboratory and modeling experiments confirmed this conclusion;
b) The proximity of multiple sources on DNAPL distribution has an important effect on overall DNAPL distribution, in that coalescence of individual plumes will result in deeper penetration into the saturated zone;
c) Comparison to both laboratory and field DNAPL flood and redistribution experiments indicates that this and other NAPL models are capable of capturing the physics of the problem. However, at the field scale, uncertainty in the most important 'driving' parameters, hydraulic conductivity and source conditions (location and release rate history), as well as, computational issues (discretization limitations), requires that DNAPL modeling be combined with groundwater quality profiling data. The profiling data provides an indication of DNAPL location, and the model is used to obtain a meaningful realization using 'effective' parameters for screening and remediation purposes.

REFERENCES

Broholm K., S.Feenstra, J. A. Cherry. 1999. "Solvent Release into a Sandy Aquifer. 1. Overview and Dissolution Behavior." *Environ. Sci. Technol. 33(5):*681-690.
Fishman M., J. Guarnaccia, C. Enfield, L. Wood. 1998."DNAPL infiltration and Distribution in Subsurface: 2D Experiment and Modeling Approach." *Nonaqueous Phase Liquids. Remediation of Chlorinated and Recalcitrant Compounds.* C1-2:37-42.
Guarnaccia, J. F., G. F. Pinder, and M. Fishman. 1997. *"NAPL: Simulator Documentation,"* EPA.   600   R-97/102, http://www.epa.gov/ada/napl_sim.html.
Kueper, B. H., D. Redman, R. C. Starr, S. Reitsma, and M. Mah. 1993. "A Field Experiment to Study the Behavior of Tetrachloroethylene Below the Water Table: Spatial Distribution of Residual and Pooled DNAPL." *Ground Water.31(5):*756-766.
Pardieck, D. and J. Guarnaccia. 1999. "Natural Attenuation of Groundwater Plume Source Zones: A Definition." *Journal of Soil Contamination, 8(1):*9-15.
Poulsen, M. M. and B. H. Kueper. 1992. "A Field Experiment to Study the Behavior of Tetrachloroethylene in Unsaturated Porous Media." *Environ. Sci. Technol.26(5):*889-895.

ACKNOWLEDGMENTS

The work upon which this paper is based was supported by U.S. Environmental Protection Agency through its Office of Research and Development under Contract Number 68-C4-0031 to Dynamac Corporation, and partially funded by the Strategic Environmental Research and Development Program (SERDP). It has not been subjected to Agency review and therefore does not necessary reflect the views of the Agency and no official endorsement should be inferred.

DNAPL – WHAT WE NEED TO KNOW AND WHY

Evan K. Nyer (ARCADIS Geraghty & Miller, Inc., Tampa, FL)

HISTORY OF DNAPL

I remember when the whole DNAPL thing began. Back then the disagreement with EPA was about the effectiveness of pump and treat to remediate chlorinated hydrocarbons. Once we realized that the pump and treat was not going to be able to clean the aquifer, the search went out to find the reason that the concentration stopped decreasing.

The federal EPA and most of writing public quickly decided that DNAPL was the one and only reason that the concentration levels would not go down during a remediation. At first, this was a good thing. The EPA decided that showing the presence of DNAPL would prove that complete cleanup to non detect was improbable, and that a new end point could be selected as the objective of the active remediation at the site. This was a major advance in the design of cost effective, technically correct, remediations.

Well now we have gone too far. It now seems that DNAPL has become the monster under the bed for the EPA and the writing public. Everything that could go wrong at a site is now blamed on DNAPL. I can see the beginnings of this phenomenon in the literature over the last several years. I personally ignored these publications as a small group of researchers that just happened to write a lot. While some have been off installing new treatment methods for chlorinated hydrocarbons, a lot of people have been concentrating on the analysis and removal of DNAPL. The problem is that these are not separate areas of interest.

Let us go through some basic truths about chlorinated hydrocarbons and DNAPL.

WHAT IS DNAPL

TCE and/or PCE are not DNAPL. DNAPL is Dense Non Aqueous Phase Liquid. This means that we have a liquid, that is not water based, and happens to be more dense then water. Everyone seems to understand this until they are asked to provide an example. The quick answer is usually TCE or PCE. However, you can not interchange the terms TCE or PCE for the term DNAPL. TCE and PCE are specific chemicals, DNAPL is the form in which a chemical or mixture of chemicals exist.

TCE can be in solution with water, adsorbed as a molecule to the soil particles in an aquifer, be a free flowing liquid, or be in solution in a liquid that is not water. Here is the test. Which of the above four categories are DNAPL? Too many of you answered the last two. The first two are not DNAPL. The free flowing liquid is DNAPL. The trick category was the last one. TCE in a non aqueous phase liquid does not make it DNAPL. Since the last example was a mixture of non aqueous phase liquids, it will depend on the combined specific gravity of the mixture. If the mixture is lighter than water than it will be a LNAPL. I am working on a project at which we have an LNAPL that is 23% PCE. Even with that high concentration of PCE, the NAPL is still lighter than water and floats on top of the groundwater. The rest of the non aqueous phase liquid is mineral oils, and the combined specific gravity is less than water.

The other way that a NAPL can exist in the vadose zone or aquifer is as a residual liquid. This means that the non aqueous phase liquid is in a small quantity and is not free flowing. The NAPL can be L or D, but will usually not move. Figures 1 and 2 show the two ways that a residual NAPL can exist below ground. While LNAPL can be readily found as a free flowing liquid or as a residual, DNAPL (if it exist at all) would usually be in the residual form. The problem with this statement is that there is no proof. I have worked on over 200 sites that have had chlorinated hydrocarbons present. None of those sites produced a free flowing liquid. Many of the sites had chlorinated concentrations above 1%, 5% and even 10% of the solubility of the particular CVOC. However, we never found any direct evidence that a residual DNAPL was present.

ALL CHLORINATED HYDROCARBONS ARE NOT DNAPL

The next test question is this. If we have 1.0 mg/l of TCE in water, and we leave the water set for a long period of time (1 month), will the TCE settle to the bottom of the water sample? Too many people think the answer to this question is yes. DNAPLs are heavier than water and will sink. A TCE/water solution is just that. The TCE molecule is part of the water solution, not independent.

Some chlorinated hydrocarbons should never be considered as possible DNAPL components. The biological breakdown products of the chlorinated hydrocarbons are not part of the original DNAPL as 1,2 DCE and Vinyl Chloride are not part of DNAPL. These compounds were never used by any company. They are the result of bacterial action on the TCE or PCE. Since bacteria only exist in water, and only interact with the chlorinated hydrocarbons that have entered the water, the TCE in the DNAPL must have solubilized in the water before the bacterial reaction took place. The TCE was no longer a separate phase liquid, it was part of the water phase.

This means that you can not use Total Chlorinated Volatile Organic Carbon (TCVOC) as a measurement to indicate the presence of DNAPL in the area. It is at this point that some people like to say that DCE and Vinyl Chloride are more soluble in DNAPL then they are in water. From this statement they make the logical leap that the presence in water would therefore mean an even greater presence in the DNAPL. The resulting conclusion is that high concentrations (see the section below for the meaning of high concentrations) of TCVOC would mean that DNAPL is in the area. Somehow this is circular logic; the presence in water proves the presence in DNAPL which proves the long term presence in water. Since these compounds first had to be in the water, the circle does not work.

My experience tells me that high concentrations of cis-1, 2 DCE and vinyl chloride means that we have reducing conditions and a highly active bacterial population. Only finding the parent chlorinated compounds are an indication of a separate phase of liquid present in the aquifer. Until someone produces some

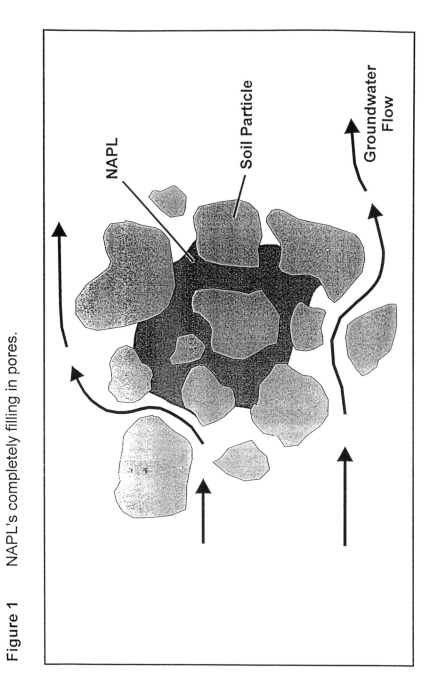

Figure 1   NAPL's completely filling in pores.

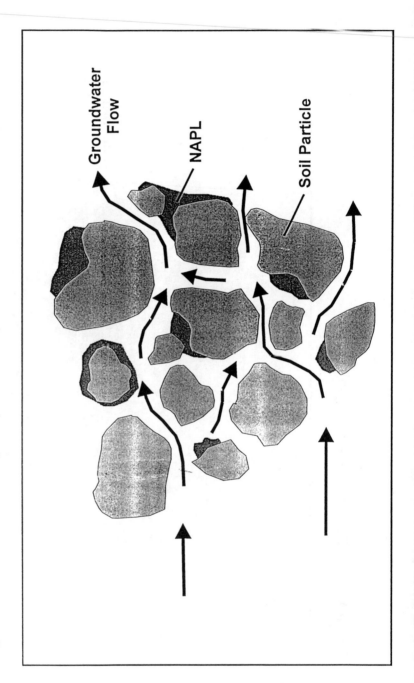

Figure 2   NAPL's coating soil particles.

compelling evidence, I suggest that we keep our DNAPL analysis to the parent chlorinated hydrocarbons: i.e. PCE, TCE, TCA.

DAPL

I could not leave this area without mentioning a Dense Aqueous Phase Liquid. Not all things that sink are DNAPL. I am currently working on a project that has a DAPL. The original material lost to the environment was a high concentration of sugars and other simple organic compounds. The concentration was high enough so that the resulting density of the liquid was greater than the groundwater. Even though the mixture was water based, it traveled through the groundwater until it was stopped by the impermeable bedrock. Many readers that have worked in the water supply area of aquifers are familiar with this concept. Aquifers along the coasts are subject to salt water intrusion when pumping an aquifer. Saltwater is denser than fresh water, and so it sinks below the fresh water. I am not sure how you are going to use this small bit of knowledge, I just thought that it was interesting.

FINDING DNAPL

The next problem is finding the DNAPL. It is very unusual to find a free flowing liquid below the water table. I would guess that less than one in one thousand (maybe ten thousand) projects have a dense, free flowing, separate phase liquid. The more normal situation is residual DNAPL. The DNAPL was originally free flowing, but as it traveled down through the aquifer soil, it was adsorbed. This left small amounts of DNAPL in the aquifer along the path of travel. This continued until all of the DNAPL was adsorbed.

The problem is that the DNAPL does not evenly coat all of the soil particles as it travels. Figure 3 shows a possible path for the free flowing DNAPL. As can be seen, the DNAPL will travel the path of least resistance. This may be a fracture, higher permeable soil, or simply the correct alignment of the soil particles. There is no way to determine the exact location of the flow path. The problem then becomes finding DNAPL in a non continuous form.

The first thought is usually to core the area and bring up the DNAPL contaminated soil. This is difficult when the DNAPL is in the saturated zone. Even with the best of conditions, this process is still hit and miss. When you have an area that may be an acre in surface area and a depth of 100 to 200 ft below ground surface, using a 2-3 in. sampling device is not representative. Finding a needle in a hay stack may be easier. What do you do with the data that you do get from this technique? If you do not find DNAPL, and the TCE in groundwater concentration is 40 mg/L which number do you rely on for your remediation design. If you do find DNAPL, and the TCE in groundwater concentration is 0.5 mg/L, which number do you rely on? In both cases, my experience says to rely on the groundwater number. If that is the case, why did we get the core data in the first place?

One method to determine the presence of DNAPL is to rely on the concentration in the groundwater. Originally, the recommendation was finding a groundwater concentration equal to or greater than 10% of the solubility of the chlorinated hydrocarbon was a strong indication of the presence of DNAPL. Over time, this was reduced to 5%, and now the recommendation is to look for 1% of the

Figure 3 Contamination plume resulting from a DNAPL spill

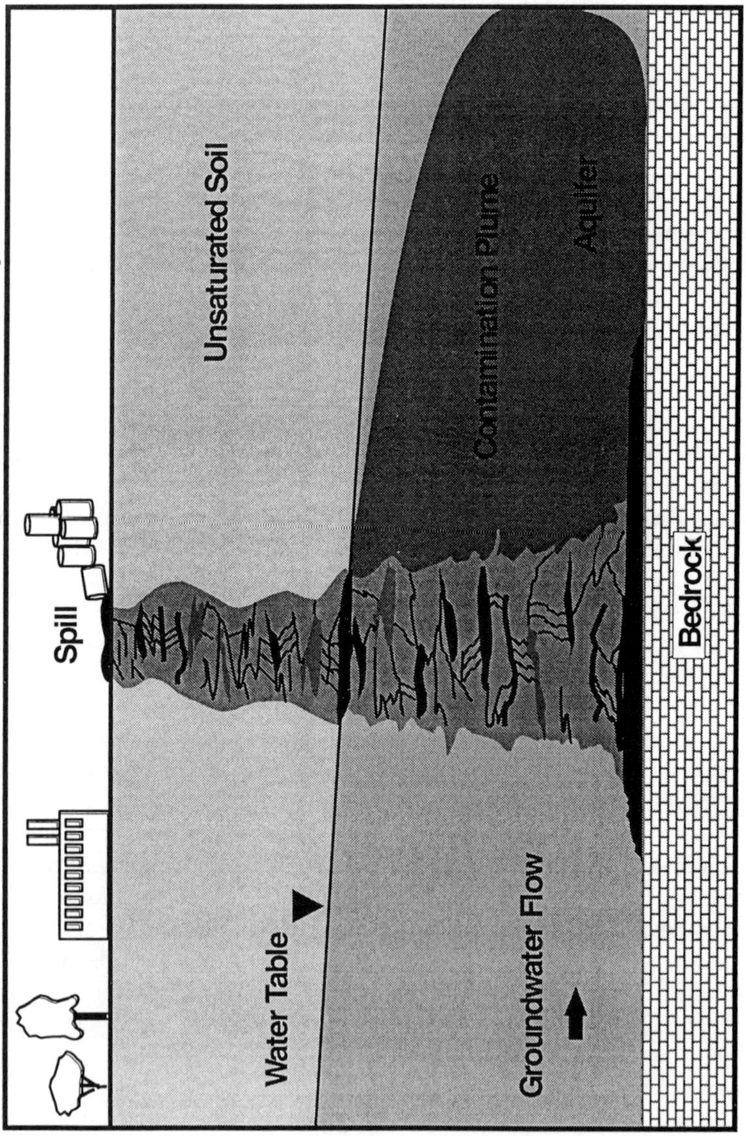

solubility.   Of course, I have been told lately that any presence of chlorinated hydrocarbon in the groundwater was a guarantee of the presence of DNAPL. Once again too many people think that this statement is true.

The other way to use the groundwater concentration is to add the time affect. This can be used before the remediation is designed or during the course of the remediation. Figure 4 shows the typical Life Cycle concentration curve for organic contaminants in groundwater.   This curve is found to describe the change in concentration while we are remediating a site or if natural attenuation is affecting the contamination before we begin active remediation. The asymptote of the curve can provide significant information on the presence of DNAPL or any other large mass of contaminants in the aquifer. The higher the concentration of the asymptote, the more mass that is present. The faster the curve reaches the asymptote, the less likely active remediation will have further affect on the concentration. While this method will not provide specific mass calculations, it is very accurate defining if there is a problem at the site. With other data collected from the site, this information can then be used to design the solution to the contaminant mass problem in the aquifer.

DNAPL HERESY

Now to peak the readers interest. IT DOES NOT MATTER IF DNAPL IS PRESENT AT A SITE. The only thing that really matters is what is the total mass of chlorinated hydrocarbons, and the relative rate of reaction from the natural attenuation. Having two thousand gallons of TCE saturated water is worse than having 1 gallon of pure TCE in the aquifer. You do the math. If that saturated water is outside the normal flow channels of the groundwater, the only way that the TCE will be removed is through diffusion. The only thing that matters is if the source area continues to discharge chlorinated hydrocarbons into the rest of the aquifer.

Remediating the TCE in either of these cases will be controlled by the geology. We can enhance the rate of removal to a certain extent, but we will not be able to develop a method that cleans the last molecule of TCE from the aquifer. The only way that we are going to be able to remove the last traces of TCE is by natural attenuation or enhance biological methods.

This means that the new DNAPL treatment technologies are going to have to co-exist with the microbiology of the site. The treatment methods can be first used to reduce the presence of chlorinated hydrocarbons.   We would use these techniques when the groundwater concentration is in the high mg/L and is not decreasing on its own. Most of the DNAPL remediation methods stop when the groundwater concentration reaches the low mg/L or high ug/L levels. At this point natural attenuation, or an enhanced bacterial method such as Reactive Zones, will have to take over to continue to reduce the chlorinated hydrocarbons. Whether or not we use pump and treat to control the source area during either of these remediations will depend on the local conditions.

There are some wild ideas about chlorinated hydrocarbons and DNAPLs circulating in the field. Some of these ideas are funny, but too may of them will distract from our goal of remediation. If we have "garbage in" in our design basis, we will have "garbage out" in our remediation design.

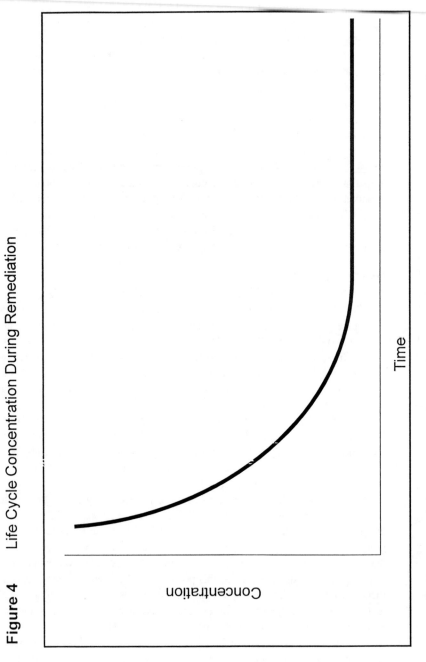

Figure 4   Life Cycle Concentration During Remediation

OVERVIEW OF IN-SITU CHEMICAL OXIDATION:
STATUS AND LESSONS LEARNED

Michael D. Basel, Ph.D., P.E. (Montgomery Watson, Des Moines, IA)
Christopher H. Nelson, P.E. (In-Situ Oxidative Technologies, Inc.,
Englewood, CO)

ABSTRACT: Lessons learned and experiences regarding the use of in-situ chemical oxidation are presented. The status of the remedial technology is summarized, including a discussion of the different chemical oxidants. Some common concerns regarding in-situ chemical oxidation are then examined, using case studies to provide insight into the merit of different concerns. Issues include safety concerns, bench-scale studies, scale-up of bench results, and oxidation in pH neutral environments. The conclusions specify other factors for which additional evaluation is needed to confirm the benefits of chemical oxidation and optimize remedial applications.

INTRODUCTION

In-situ chemical oxidation is an innovative technology that has gained widespread consideration for remediation of organic contaminants in soil and groundwater. The technology is promising because it is rapid, low cost, and relatively unaffected by contaminant characteristics and concentration. However, the reported effectiveness of chemical oxidation has been widely varied; some outstanding successes, some disappointing and outright-dangerous failures, and all variations in between. Compilations of results from previous oxidation applications have been provided in numerous publications, including the following:

- In Situ Remediation Technology: In Situ Chemical Oxidation. Office of Solid Waste and Emergency Response (OSWER). September 1998.
- In Situ Chemical Treatment. Ground-Water Remediation Technologies Analysis Center (GWRTAC). July 1999.
- Technology Status Review: In Situ Oxidation. Environmental Security Technology Certification Program (ESTCP). November 1999.

As an innovative technology, in-situ chemical oxidation faces many of the typical obstacles to widespread application. These obstacles, both real and perceived, include technical uncertainties, confusion regarding performance, and application by inexperienced personnel. This paper evaluates some common concerns and questions that often arise when discussing the merits of chemical oxidation. Case studies and personal experience are used to provide insight into the merit of different concerns. Therefore, this evaluation will help focus future efforts on the remaining uncertainties and barriers to effective implementation.

OVERVIEW OF IN-SITU CHEMICAL OXIDATION

In-situ chemical oxidation involves the injection of an oxidant to transform contaminants into relatively benign by-products such as carbon dioxide and water. Most applications have focused on using oxidants to chemically transform contamination, although selected oxidants can be used to enhance bioremediation by making some contaminants more available for bacterial metabolism and by providing electron acceptors such as oxygen. Chemical oxidation is particularly favorable for chlorinated compounds because the process does not produce vinyl chloride, a common biodegradation product.

The oxidants most frequently used are Fenton's reagent (hydrogen peroxide and iron), potassium permanganate, and ozone. A summary of these oxidants, including oxidation potential, by-products, reaction times, and costs, are shown in Table 1.

TABLE 1. Comparison of Oxidants

	Fenton's Reagent	Permanganate	Ozone
Physical State	Liquid	Liquid	Gas
Molecular Composition	$OH^{\bullet}$	MnO_4^-	O_3
Hydroxyl Radical Formation	Yes	Under very limited conditions.	Under certain conditions.
Oxidation Potential	2.80 V (highest)	1.70 V	2.07 V
Reaction Times	Very Fast	Slow	Fast
Contaminant Treatability	Many organic contaminants	Few organic contaminants	Some organic contaminants
By-Products	Ferric iron, oxygen, and water.	Dissolved manganese, manganese dioxide, and potential heavy metals.	Oxygen
Potential to Enhance Bioremediation	Yes, if applied under neutral conditions.	Unlikely	Yes
Capital Costs	Low	Low	High
Reagent Costs	Moderate	Moderate	Moderate
Metals Mobilization/Oxidation Potential	May cause metals mobilization if applied under acidic conditions. Insignificant measured chromium oxidation under neutral conditions.	Permanganate and manganese dioxide can oxidize Cr^{3+} to Cr^{6+}. Increased cation/anion displacement potential from soil matrix due to potassium interactions.	Metals oxidation potential.

Hydrogen peroxide may be injected alone or in conjunction with an iron catalyst, which would utilize Fenton's chemistry to form hydroxyl radicals. The use of Fenton's chemistry to form a hydroxyl radical rather than straight injection of hydrogen peroxide is more favorable because the hydroxyl radical is a stronger oxidant. Fenton's reactions are relatively inexpensive and fast acting, but require a catalyst and may require pH adjustment to acidic conditions for classic Fenton's chemistry, although research has indicated Fenton-like reactions can occur under neutral pH conditions (Watts and Stanton, 1994).

Potassium permanganate ($KMnO_4$) oxidizes contaminants directly without catalyst or requirements for pH control. Since potassium permanganate is not as strong an oxidant as hydroxyl radicals, more potassium permanganate must be injected but permanganate has the potential to migrate further into the formation before being consumed. However, permanganate is more expensive, produces manganese oxide precipitates, increases dissolved manganese levels, and can cause water to turn purple if all permanganate is not consumed.

Ozone (O_3) achieves direct oxidation using injection of gaseous O_3. Ozone may be unstable and is more suitable for treatment of contaminants, such as petroleum hydrocarbons, for which the breakdown of additional oxygen would enhance aerobic degradation. Since it is a gas, ozone can be dispersed further in the vadose zone than the other liquid oxidants.

LESSONS LEARNED

Risks of In-Situ Chemical Oxidation Can Be Safely Managed. Two common safety concerns regarding in-situ chemical oxidation are the potential risks caused by use of hazardous chemicals and the possibility of uncontrolled reactions. These concerns are certainly valid, but experienced staff following proper procedures can mitigate these risks.

Because strong oxidants are highly corrosive and potentially explosive, care should be taken during field activities. Of course, the actual risk posed by strong oxidants depends on the concentrations in solution. Using hydrogen peroxide as an example, bottles of hydrogen peroxide are commonly found in households at 3% concentration for use in sterilizing open cuts or even as mouthwash. Concentrations of hydrogen peroxide used for remediation are typically higher, ranging from 8% to 50%. At the low end of this range, the aqueous solution poses little more risk than the bottle in household medicine cabinets. At the high end, much more care should be taken when handling and injecting the solution, but this risk can also be managed if proper health and safety plans are prepared and followed. A summary of the methods that can be utilized to manage the safety concerns are summarized in Table 2.

Table 2. Procedures to Manage Safety Concerns Associated with In-Situ Chemical Oxidation

- Work should be performed by experienced personnel
- Proper procedures should be followed at all times
- Work should not be rushed
- Oxidants should be at the lowest appropriate concentration
- Dermal contact should be prevented with proper clothing, including eye shields
- Inhalation should be minimized by performing work in well-ventilated areas
- In-situ mass transport should be understood and controlled

Safety concerns have also focused on reports of uncontrolled reactions in the subsurface (Nyer and Vance, 1999). Chemical oxidation is an exothermic reaction generating heat that can increase temperature and pressurize gases depending on loading rates and reaction rates. Again, these risks can be mitigated through the use of experienced professionals to plan and implement all field activities. Site conditions that would warrant particular attention in the planning stage include paved sites in which vapor pressures could buildup under the cap, sites with preferential flow paths or utility corridors through which vapors could migrate, and beneath buildings. Simple strategies that can be used to mitigate risks at such sites is the use of lower concentrations of oxidants, lower reagent volumes, and lower injection pressures.

Effective, Safe Mass Transport of Reagents is Crucial for Success in the Field. For remediation to be effective, injected oxidants have to contact a large percentage of the subsurface contaminants. Otherwise, the bulk of oxidant/reagent will be consumed by natural organics with significant contamination remaining behind in soil and groundwater. Effective distribution of reagents require a suitable delivery mechanism and a good understanding of subsurface conditions that affect mass transport. Therefore, a lesson learned from published reports of successful and unsuccessful case studies is that in-situ chemical oxidation must be performed by professionals experienced with the technology and a good understanding of the site geology and hydrogeology.

Bench-Scale Tests Provide Basis for Field Applications. A common question asked when in-situ chemical oxidation is being considered is whether bench-scale tests are really necessary. Conventional wisdom has historically recommended bench-scale testing because the tests were a relatively inexpensive way to demonstrate effectiveness, at least under ideal conditions, of the innovative technology. Bench scale tests provide the most value in demonstrating the oxidation chemistry, estimating the overall oxidant demand of the aquifer matrix, and producing a basis for regulators (and clients) to believe that in situ oxidation will work at their site. Once the oxidation chemistry is demonstrated and the oxidant demand is estimated, a field pilot test should be performed to focus on mass transport issues, i.e., can the oxidant reagents be effectively and safely distributed in the subsurface at the site.

A bench-scale test would not be required if the can be predicted with confidence without actual testing. As a minimum guideline, the following requirements would need to be satisfied before skipping the bench test:

✓ Chemical constituents are well characterized.
✓ Oxidant reagents have been proven effective for site contaminants.
✓ Oxidant demand is understood.
✓ Optimal reagent formulation can be estimated.
✓ Stakeholders all have confidence in technology.

However, even if all the above conditions are met, bench-scale tests can still provide useful benefits. Table 3 summarizes the benefit of a bench-scale test for a site where conditions were supposedly well understood and were well suited for oxidation. In this case, the bench-scale test demonstrated the presence of unexpected oxidant demand, which lowered the effectiveness of treatment. As a result of the bench test, stakeholder expectations were adjusted such that the field application could meet expectations.

Table 3. Case Study Demonstrating Benefit of Bench-Scale Tests

Subsurface Conditions	Contamination present in sand bedding immediately beneath floor of building. Clay soils beneath sand. Water table at 7' bgs not impacted.
Contamination	Petroleum hydrocarbons with approximate extent of 4000 sq.ft. Facility initially indicated no other releases occurred.
Results of Bench Scale Test	Reduction of contaminant under ideal conditions in lab only 44%, which was lower than expected.
Reason for Lower Effectiveness	Facility subsequently discovered that fire-fighting foam had been released at site. Foam would provide significant carbon source consuming oxidant.
Benefit of Bench Scale Test	Expectations were re-adjusted based on bench-scale test. Full-scale application proceeded with regulatory approval with anticipated cost savings of 50% compared to alternatives.

Typical Results of Successful Bench-Scale Tests. Bench scale tests typically demonstrate high effectiveness (90 to 99% mass reduction) for oxidizable compounds such as TCE and petroleum hydrocarbons. If the subsurface chemistry is not favorable to chemical oxidation, mass reduction will be significantly lower, as was demonstrated in the example in Table 3. An overview of representative mass reductions achieved in bench scale studies for TCE and petroleum hydrocarbons, for favorable sites, is shown in Figure 1.

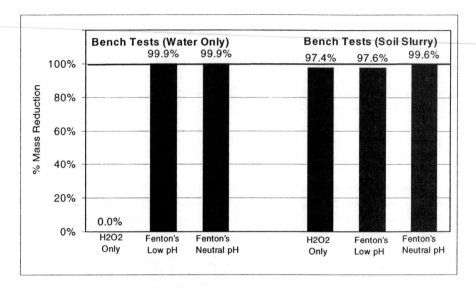

FIGURE 1. TCE Mass Reductions Achieved in Bench Scale Tests

Fenton-like Chemistry at Neutral pH. The use of hydrogen peroxide with iron catalyst to oxidize contaminants via Fenton's reactions is the most widespread application of chemical oxidation. In addition to classic Fenton's chemistry, there has been reported success using "Fenton's-like" chemistry that utilizes chelated metal catalysts and hydrogen peroxide under neutral pH conditions. This is a significant development in that most contaminants are found in neutral pH environments.

ISOTEC has developed a process that can sustain Fenton's oxidation and the production of hydroxyl radicals (Greenberg et al., 1998). Comparison of bench-scale studies for pH-neutral and acidic formulations as well as results of actual field applications of pH-neutral chemical oxidation have demonstrated successful applications. Figure 1 provided a comparison of bench scale results for Fenton's chemistry (or Fenton-like chemistry) for oxidation of TCE using catalysts formulated for low pH and neutral pH environments. The bench-scale tests indicates that results for the different catalysts are comparable.

A summary of bench-scale and field scale mass reductions that have been achieved using neutral pH catalyst is provided in Table 4. Further details for these case studies are readily available upon request. This table demonstrates that in-situ chemical oxidation can be successful in the field under neutral pH conditions.

Table 4. Representative Case Studies For In-Situ Chemical Oxidation Utilizing Fenton-like Chemistry Under Neutral pH Conditions

Contaminants	% Mass Reduction	
	Bench Tests	Field Results
Benzene, Toluene, Ethylbenzene, Xylenes (BTEX)	99.9%	98%
BTEX	99.9%	99%
BTEX	94%	72 – 93%
Ethylbenzene, Xylenes / Trichloroethene, cis-1,2-dichloroethene, Vinyl Chloride	99%	99%
Perchloroethene, Trichloroethene	97%	70%
Perchloroethene	93%	27% - 74%
Perchloroethene	99%	76%
Polyaromatic Hydrocarbons (i.e. Napthalene)	94%	78%

CONCLUSIONS

In-situ chemical oxidation is an innovative remedial technology with widely varying opinions regarding its future effectiveness for clean up of contaminated soils and groundwater. As additional work is performed to evaluate chemical oxidation, it will be imperative that the work be focused on true limitations and opportunities of the technology. This paper has evaluated several common concerns and provided data to these concerns. Concerns that were addressed in this paper and determined to be minor include: safety concerns, bench-scale studies, scale-up of bench results, and oxidation requiring acidic pH environments.

This leaves several true limitations that merit further consideration by the professional community, including researchers, academicians, stakeholders, and remediation specialists. Further consideration of these factors is needed for two primary reasons: 1) Continued confirmation of the true effectiveness of chemical oxidation as a worthy remedial technology and 2) Identification of the proper characteristics needed to facilitate optimal use of chemical oxidation. Some of the most important factors include the following:

- Mass transport and contaminant contact with oxidants in the subsurface
- Monitoring of potential groundwater rebound over time
- Effect of oxidation on future biodegradation
- Increased understanding of mass transport issues, such as distribution of oxidant from injection point, longevity of oxidant in subsurface, and impact of natural oxidant demand
- Compatibility with other remedial technologies
- Field implementation methods

REFERENCES

Greenberg, R.S., Andrews, T., Kakarla, P.K.C., and Watts, R.J.., 1998. "In-Situ Fenton-Like Oxidation of Volatile Organics: Laboratory, Pilot and Full-Scale Demonstrations", *Remediation*. Spring.

Nyer, Evan and David Vance, 1999. "Hydrogen Peroxide Treatment: The Good, the Bad, the Ugly." *Ground Water Monitoring Review*. Summer: 54-57.

Watts, R.J. and P.C. Stanton, 1994. "Process Conditions for the Total Oxidation of Hydrocarbons in the Catalyzed Hydrogen Peroxide Treatment of Contaminated Soils." *WARD 337.1*, Washington State Department of Transportation, Olympia, Washington.

ON SOURCE-ZONE FLOODING FOR TREATING DNAPL SITES

Motomu Ibaraki (The Ohio State University, Columbus, Ohio)
Franklin W. Schwartz (The Ohio State University, Columbus, Ohio)

ABSTRACT: Several technologies for cleaning up DNAPLs in source zones rely on schemes that solubilize contaminants or destroy them *in situ*. Typically, these approaches employ an injection/withdrawal system to recirculate the treatment fluids. This study looks at the combined influence of gravity and porous medium heterogeneities and the analysis is based on a series of numerical simulations. Results indicate that the characteristics of convective mixing greatly affect removal efficiency and displacement patterns. The ratio of Grashof number and Reynolds number proved useful in interpreting the patterns of flooding in the homogeneous porous media. When higher permeability layers are included in the domain, they act as conduits and produce significant flooding inefficiencies. The problem is less severe in heterogeneous media where the connectivity through the treatment zone is less well developed. Overall, this paper illustrates that density effects and the potential development of high permeability pathways need to be considered in designing chemical floods.

INTRODUCTION

Several remedial technologies for the treatment of DNAPL source zones rely on schemes that solubilize contaminants or destroy them *in situ*. Examples include; alcohol flooding, surfactant flooding, or permanganate flooding. Commonly, these approaches are employed in field settings with an injection/withdrawal system. Wells are situated to maximize the volume of treatment fluid passed through the zone of contamination.

As Jawitz et al. (1998) point out, much of the initial research on these types of schemes has focused on extraction efficiencies or how well the contaminants are removed or destroyed by the flushing fluids. Much less work has been concerned with the flooding behavior. In the petroleum industry, flooding behavior is characterized by the so-called sweep efficiency. It is the total volume of zone of contamination contacted during the flushing process. The most efficient flushing comes from a stable piston-like displacement. Low sweep efficiencies can develop as a consequence of porous medium heterogeneities that channelize flow (e.g., Walker et al., 1998) and unstable displacements, caused by viscous fingering, and gravity underrides and overrides (Jawitz et al., 1998).

The purpose of this paper is to examine some factors that influence the ability of floods to remove DNAPLs especially in source zones. Our particular emphasis is on the combined influence of gravity and porous medium heterogeneities. The approach involves simulations that are highly resolved in space and time with a simulation scale that is typical of field applications. This work represents an initial step toward our longer-term goal of improving removal efficiencies.

Methodology

Study Design. The study design involves preparing and testing prototypical systems that encompass varying patterns of media heterogeneity and densities of the treatment fluids. The design of these simulation trials is discussed in the following section. Another issue of study design is how to interpret the results of the simulation trials in a way that is appropriate for an overall assessment of flooding success. One possible way is to determine the sweep efficiency for the given trial. Sweep efficiency is defined as the fraction of the zone of contamination influenced by the flushing process. In problems of oil recovery, the concept of sweep efficiencies is a good one because the wells are often far apart and problems, like unfavorable mobility ratios, can cause large segments of a reservoir to be by–passed. With contamination problems, injection/withdrawal wells are placed closer together and it is more difficult (but not impossible) to sweep the DNAPL source zone in an inefficient manner.

The concept of sweep efficiency is useful in evaluating the extent to which a unit is either flooded or not flooded with some treatment chemical. In con–taminant applications, the quantity of treatment chemi–cal being delivered to vari–ous points in the zone being swept is also of importance. For example, if some vol–ume of porous medium only receives a small flux of per–manganate only a corre–spondingly small quantity of contaminant can be de–stroyed. Similarly, small fluxes of a solubilizing sur–factant will produce slow rates of contaminant disso–lution. A high sweep effi–ciency is a necessary but not sufficient condition to ensure source–zone cleanup. Thus, the analysis here also evalu–ates the theoretical "treat–ment flux ratio" of all the model cells in the zone of contamination. The treat–ment flux ratio is defined as the cumulative magnitude of advective flux of the treat–ment chemical injected that reaches points of interest in the zone of contamination compared to that at the in–

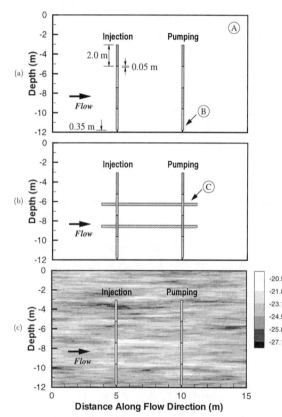

FIGURE 1. Schematic representation of chemi–cal flushing scheme and model domain.

jection wells. Because the treatment flux ratio is defined at various points, con–toured cross–sections will illustrate the potential for various parts of the contami–nated zone to be remediated. We use the term ''theoretical'' because this analysis does not account for the utilization of the treatment chemicals. In reality, the downstream end of the contaminated zone will need to wait its turn actually to receive the treatment chemical. However, this approach reflects the local inten–sity of flooding of the treatment chemical.

Description of Simulation Experiments. The simulation experiments are conducted within a 2–D region, 15 m long and 12 m high. Wells are used to create an injection/withdrawal doublet to force flow through a treatment zone between the wells (Figure 1a). Given the symmetry of the system, the pattern of injection/withdrawal provides infinite lines of wells. The injection and withdrawal wells are installed with four separate screen sections that are each 2 m long. This design feature attempts to reduce the effects of preferential flow caused by heterogeneities in hydraulic conductivity.

The pumping is superimposed on a horizontal, ambient flow that is moving from left to right (Figure 1a). This flow condition is developed using no–flow boundaries along the top and bottom of the domain, a flux boundary along the left side and right sides, with a constant pressure (100 Kpa) reference node located at the top left corner. In all cases, the ambient ground water velocity is 30 cm/day. With the wells in operation, the velocity of fluid flow between the injection/withdrawal wells increased to 45 cm/day.

TABLE 1. Parameter values for numerical simulations.

Parameter	Value
pcrmcablity	
Layer A	5×10^{-11} m^2
Layer B	2×10^{-7} m^2
Layer C	5×10^{-10} m^2
porosity	0.25
free–solution diffusion coefficient	1.61×10^{-9} m^{2}s^{-1}
longitudinal dispersivity	1.0×10^{-3} m
transverse dispersivity	1.0×10^{-4} m
tortuosity	0.35
fluid discharge	2.19 ml/min
Mean permeability	5.0×10^{-11} m^2
Longitudinal correlation length	1.8 m
Transverse correlation length	0.15 m
Variance of the log–transformed permeability field	0.9

Three different media are utilized for these simulations. Medium A is a non–layered homogeneous and isotropic medium having a permeability of 5.0×10^{-11} m². Medium B has two, thin, permeable layers that extend between the wells (Figure 1b). The background permeability is 5.0×10^{-11} m² and the layer permeability is 2.85×10^{-9} m². Medium B (Figure 1c) is a correlatedrandom permeability field generated using the approach of Robin et al. (1993). The geostatistical properties of this field are provided in Table 1. A permeability field with a variance of 0.90 is quite heterogeneous as compared to sites like Canadian Forces Base Borden and Cape Cod. The complete specification of this simulation problem requires other hydraulic and mass transport parameters. Details are provided in Table 1. For this relatively small system, dispersivities are assumed to be at the low end of the range for natural materials (Domenico and Schwartz, 1998). The model accounts for the variation in density and viscosity according the data provide in Table 1.

The mass transport model also requires boundary and initial conditions. The top, bottom, and left side boundaries are assumed to be Type 2 (i.e., no dispersive flux). The right side boundary is assumed to be a free–exit boundary. Mass loading at the injection wells is represented by a constant concentration boundary. A specified–flux boundary prescribes the flow. The withdrawal wells are incorporated as free–exit boundaries for solute and specified–flux boundaries for flow. The initial condition is that $C(x,z,0) = 0$ g/L.

Flow and mass transport are simulated using the flow and transport code MITSU3D (Ibaraki, 1998). This code is capable of solving variable–density flow and transport problems in two or three dimensions with up to seven million nodes. The variable–density formulation requires the flow and transport equations be fully coupled. The large number of nodes, the small time steps and inherent non–linearity requires that the solver be both efficient and robust.

INTERPRETATIONS

The evaluation of flooding behavior is conducted using concepts of sweep efficiency and treatment flux ratio that were developed earlier. It turns out that sweep efficiency is most useful as a concept applied to the analyses with Medium A, because changing the fluid density produces significant changes in sweep efficiencies. The heterogeneous media (B and C) are examples where sweep efficiency as a concept is not particularly useful. The pattern of flooding is such that even with differences in fluid density the treatment fluid will flush the contaminated zone efficiently. In other words the sweep efficiency with respect to the contaminated zone is excellent. However, for Media B and C, the delivery efficiency will be relatively poor for large portions of the contaminated zone.

Flooding Behavior with the Homogeneous Medium (A). The simulation trials with the Medium A showed that increasing the density contrast between the treatment fluid and the ambient fluid had a marked influence on the behavior of the flood. These simulation trials reflect a transition in behavior from a purely forced convective flow system to a mixed convective system where both forced and free convection operate to transport mass. In a forced convective system, dissolved mass is transported by the flowing ground water but does not influence the driving forces for flow. In free convection, fluid flow is driven by differences in fluid density.

Dimensionless numbers provide a convenient way to evaluate the extent to which these various driving forces are operative. The strength of forced flow is often characterized by Reynolds number (*Re*). In contrast, the "vigor" of buoyancy–induced flow is characterized by Grashof number (*Gr*). The ratio of these parameter gives us relative vigor of forced to free convection (Gebhart et al., 1988). When *Gr/Re* becomes infinite, the flow is density driven. When *Gr/Re* approaches zero, the transport is due to forced convection. For simple homogeneous systems, one can apply these equations to determine what kind of system is likely to develop in a given flooding scheme.

In order to examine the effects of convection process on sweep efficiency, we conducted 17 simulations with Medium A in which permeability and injected fluid concentration are varied in the range of 2.9×10^{-11} to $6.0 \times 10^{-13} \, m^2$ and 0 to 4.0 %, respectively. All of the simulations are continued until one pore volume of fluid is added (based on the porous medium between the wells) or the treatment fluid reached the left side boundary, whichever is first. We needed this second criterion because treatment fluid will be lost to outside of domain after this time. Also, treatment fluid moving left of the injection well, i.e., up gradient, is considered to be wasted. We define the ratio of the quantity of mass moving upstream and the total mass injected in the system as the waste ratio.

Figure 2 describes the relationship between the sweep efficiency or waste ratio and *Gr/Re*. As *Gr/Re* increases, i.e., the influence of free con–vection becomes more dominant, the early–time sweep efficiency is about 50%. Once *Gr/Re* increases to above about 3, there is a significant quantity of treatment fluid wasted by up stream migration. The waste ratio then gradu–ally increases to 45% as *Gr/Re* increases.

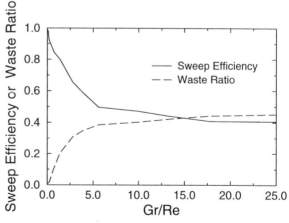

FIGURE 2. The relationship between the sweep efficiency or waste ratio and *Gr/Re*.

Calculating the dimensionless *Gr/Re* ratio for a problem involved with flooding of a source zone with a treatment fluid should provide an indication of potential inefficiencies related to density driven flow. These dimensionless numbers are functions of the properties of the various fluids and the porous medium. As Figure 2 suggests, sweep efficiencies for a homogeneous medium may begin to suffer once *Gr/Re* becomes larger than 3.

Flooding Behavior with the Heterogeneous Media (B, C). The more important question that we alluded to previously is whether the flooding will do any good in a remediation context. For example, in a reasonable period of time surfactant flooding or oxidation by permanganate will require that many pore volumes of the treatment fluid, i.e., high magnitude of advective flux, pass through the contaminated zone. It is evident with Medium B that the intensity of flooding is extremely variable. Large quantities of the treatment fluid are being moved through the two high permeability layers. The lower permeability domains receive much less volume of treatment fluid although they are being swept efficiently. This extreme variability in the local mass flux of treatment fluids moving through the contaminated zone does not occur in homogeneous domain because uniformly distributed flow paths and relatively small variation in the magnitude of advective velocities.

For the simulation trials with the heterogeneous medium, we calculated the treatment flux ratio or the ratio of flux magnitude of treatment fluid passing through various locations in the contaminated zone compared to that at the injection wells. Not surprisingly, with Medium B, the two high permeability zones receive more than 10 times higher amount of treatment chemicals as compared to 0.04 to 0.1 over the remainder of the contaminated zone. Looking in particular at zones immediately adjacent to the injection well, there is significant variability the treatment flux ratio (Figure 3). This heterogeneous pattern is caused by free convective that develops at this point in the system.

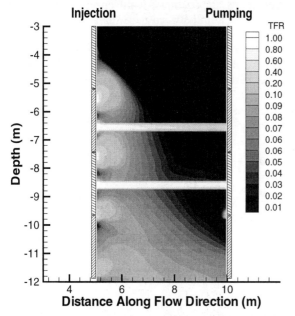

FIGURE 3. Treatment Flux Ratio (TFR) distribution for the two-layer domain case.

With Medium C, permeable pathways are somewhat more difficult to see (Figure 1c) but nevertheless they do indeed exist. At least two zones promote higher mass fluxes of the treatment fluid through the system. As Figure 4 shows, the treatment flux ratio is high (> 1 treatment flux ratio) along these zones at early time. Other zones, for example the intervals from 10 to 12 m and 5 to 7 m have much lower delivery efficiencies in a range of 0.01 to 0.04. Again, the treatment chemical tends to migrate through the high permeable zones, as compared to the reset of the domain. If the remediation scheme requires that high amount of the treatment chemical be delivered, then most of the clean up will take place within the high permeability zones. In contrast, if only little amount of the treatment chemicals are required for cleanup then there is a better chance of affecting a cleanup. Nevertheless, the higher permeability zones are a constant short circuit for treatment chemicals. One might incur increased costs of disposing of effluent at the withdrawal well or a significant treatment and recycling operation.

The main problem with the heterogeneous medium is that direction of flooding coincides with the orientation of high permeability zones. These zones provide significant pathways for flow of the treatment fluid at the expense of lower permeability zones. So far, our analysis has considered heterogeneities in permeability and density. There are other potential problems that in our opinion can exacerbate the problem of delivery efficiency. Original heterogeneity in DNAPL saturation coupled with multiphase effects on permeability is probably the most important. We expect that more permeable zones will have lower DNAPL saturations than less permeable zones. This pattern of saturations would promote even more flow of treatment chemicals through the higher permeability zones and less zones flow through the low permeability zones. Obviously, this issue is important for future work.

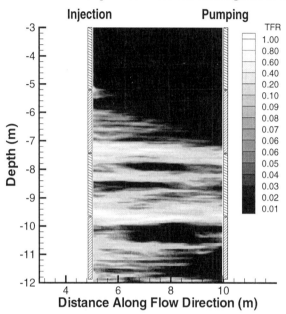

FIGURE 4. Treatment Flux Ratio (TFR) distribution for the correlated random permeability field.

SUMMARY AND CONCLUSIONS

This paper showed that density effects and heterogeneity in permeability influences the behavior of chemical floods, which are designed to remediate source zones contaminated with DNAPLs. When there is a significant difference in density between the treatment and ambient flows there is a possibility of mixed convection, where fluid convection plays an important role in fluid flow. In homogeneous porous media, density driven flow makes it more difficult to control where the treatment fluid is going. The most effective flushing comes from a stable piston–like displacement, which occurs with forced convection.

The ratio of the Grashof number (*Gr*) to Reynolds number (*Re*), Gr/Re, provides a quantitative indication of the vigor of free versus forced convection. When this ratio is above 3 in a homogeneous medium, there is a deleterious impact on sweep efficiency. Density driving forces even cause flow of the treatment fluid upgradient. This fluid misses the zone of contamination and is essentially wasted.

With heterogeneous media, high permeability pathways act as hydraulic short circuits and facilitate the flow of treatment fluids from the injection to the withdrawal wells. Density effects are less important with forced convection through the most permeable layers. Not surprisingly, the permeable pathways attract flow from the zones of lower permeability. Overall, the pattern of flushing is extremely uneven with the permeable zones being flushed by 5 times the magnitude of treatment chemical as the lower permeability zones.

The simulation results presented here make it clear that there can be real problems in flooding a zone of contamination in the direction of high permeability trends. Even quite modest pathways can end up carrying a significant flow of the treatment fluid to the detriment of flushing in the lower permeability zones. Preliminary indications are that it will be difficult to achieve the delivery efficiencies commensurate with a 95%+ clean up for flooding schemes of the type that we tested.

REFERENCES

Domenico, P. A. and F. W. Schwartz (1998). *Physical and chemical hydrogeology*. John Wiley & Sons, Inc. New York.

Gebhart, B., Y. Jaluria, R. L. Mahajan, and B. Sammakia (1988). *Buoyancy–induced flows and transport*. Hemisphere.

Ibaraki, M. (1998). "A robust and efficient numerical model for analyses of density–dependent flow in porous media." *Journal of Contaminant Hydrology*. 34 (3): 235–246.

Jawitz, J. W., M. D. Annable, P. S. C. Rao, and R. D. Rhue (1998). "Field implementation of a Winsor type I surfactant/alcohol mixture for in situ solubilization of a complex LNAPL as a single phase microemulsion." *Environmental Science &Technology*. 32(4): 523–530.

Robin, M. J. L., A. L. Gutjahr, E. A. Sudicky, and J. L. Wilson (1993). "Cross–correlated random field generation with the direct Fourier transform method." *Water Resources Research*. 29(7): 2385–2397.

Walker, R. C., C. Hofstee, J. H. Dane, and W. E. Hill (1998). "Surfactant enhanced removal of PCE in a nominally two–dimensional, saturated, stratified porous medium." *Journal of Contaminant Hydrology*. 34 (1–2): 17–30.

DNAPL REMEDIATION UTILIZING CONTAINMENT TECHNOLOGIES AT A LANDFILL SITE

Frederick W. Blickle, III, (Conestoga-Rovers & Associates, Romulus, Michigan)
Joe Medved, (General Motors Corporation, Detroit, Michigan)
Michael D. Miller, (Harding Lawson Associates, Farmington Hills, Michigan)

ABSTRACT: The Middleground Landfill Site is situated on the west side of Middleground Island, which is bounded by the Saginaw River East and West Channels in Bay City, Bay County, Michigan. The Monitoring Well 8 (MW-8) Area lies on the west side of the landfill, approximately 16 m east of the Saginaw River West Channel, which presented a challenge during the remedial activities at the Site. The subsurface geology consists of an upper clay capping layer (Unit 1) that ranges from several centimeters to 7.5 m in thickness, followed by 3 to 5 cm of sand and refuse materials (Unit 2). Unit 3 is 7- to 15-m thick and consists of silty clay with a permeability of 2.3×10^{-7} cm/s (Unit 3). Dense non-aqueous phase liquid (DNAPL) materials containing polychlorinated biphenyls (PCB), located with Unit 2, were estimated to have a volume of approximately 5,400 L with an aerial extent of approximately 350 m^2. In response to the determinations of the United States Environmental Protection Agency, Michigan Department of Environmental Quality, and as required by the Michigan Act 451 Part 201 Section 14(1)(f) regarding the presence of free product, an interim response action (IRA) for the MW-8 Area was completed.

The IRA was a combination of containment and immobilization, which utilized watertight sheet piling, soil permeation grouting, and capping. The sheet pile was driven into the top 1.5 m of Unit 3 around the MW-8 Area to create a 25 m(80 feet) by 30 m(100 feet) rectangular enclosure that completely contained the DNAPL materials. Landfill objects, such as large logs, presented difficulties during the sheeting installation. The objective of the soil permeation grouting was to fill the effective porosity with grout and cut off potential groundwater flow through Unit 2, which would stop any movement of the DNAPL materials. A cap was installed to prevent moisture from entering the MW-8 Area. Current groundwater monitoring demonstrates that the IRA is effectively containing and immobilizing DNAPL within the MW-8 Area.

INTRODUCTION

This document discusses the interim response actions (IRA) completed at the Monitoring Well 8 (MW-8) Area located at the Middleground Landfill Site (Site). The IRA was completed to contain and immobilize DNAPL that occurred at the base of a thin water bearing unit. The Site is located at 1112 Evergreen in Bay City, Michigan. Construction activities were complete in the fall of 1998 at a cost of approximately $1.4 million U.S.

Objective. The objective of the soil permeation grouting was to fill the effective porosity with grout and cut off potential groundwater flow through Unit 2, which would stop any movement of the DNAPL materials. In addition, it was desirable to minimize waste material generation during this process so material would not have to be transported off Site for disposal.

Approach. The approach of this IRA was to first install sheet pile and then perform permeation grouting within the sheet piled area. Permeation grouting included laboratory determination of the grout mixture parameters, treatment of groundwater collected during dewatering of the MW-8 Area for use in the grout mixture, determination of the optimal spacing for grout injection boreholes through a Field Test Program, and then implementation of full-scale grouting. During full-scale operation, grout was injected at each borehole until the terminal grouting pressure was reached, which was a target pressure of 3.5 KN/m^2 per 0.31 m (½ pound per square inch (psi) per foot) of borehole below land surface (bls).

Figure 1 illustrates the borehole configuration for the Field Test Program and for full-scale grouting.

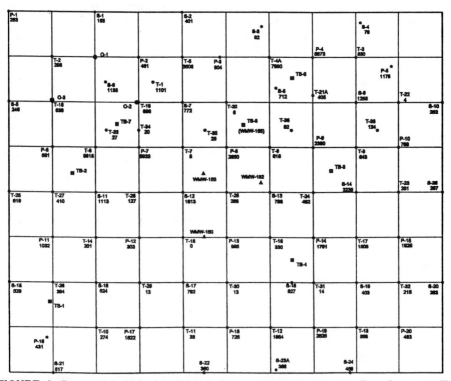

FIGURE 1. Grout borehole locations within sheet-piled area. P-•, S-•, and T-• **Indicate locations for primary, secondary and tertiary boreholes, respectively. The volume of grout is shown adjacent to the borehole. Monitoring and observation wells are shown as squares or triangles.**

Site Description. The Middleground Landfill Site is situated on the west side of Middleground Island, which is bounded by the Saginaw River East and West Channels in Bay City, Bay County, Michigan. The MW-8 Area lies on the west side of the landfill, approximately 16 m east of the Saginaw River West Channel. The island and landfill were created when Saginaw River materials were deposited into the former wetland area.

The subsurface geology consists of an upper clay capping layer (Unit 1) that ranges from several inches to 7.5 m in thickness, followed by 3 to 5 m of sand and refuse materials (Unit 2). Unit 3 is 7 - to 15-m thick and consists of silty clay with a permeability of 2.3×10^{-7} cm/s (Unit 3). Dense non-aqueous phase liquid (DNAPL) materials containing polychlorinated biphenyls (PCB), located within Unit 2, were estimated to have a volume of approximately 5,400 L with an aerial extent of approximately 350 m^2.

METHODS AND MATERIALS

Treatability Studies. The grout treatability studies determined that a bentonite-based grout would provide suitable stabilization, was compatible with the waste material and exhibited no significant leaching. During field grouting, the bentonite-grout-water mix ratio was adjusted in order to maximize grout penetration. The findings from the water treatment treatability were used to determine the appropriate treatment components, which included separation, aeration and carbon polishing.

Health and Safety. Health and Safety procedures were maintained and air monitoring was conducted throughout operations to ensure the protection of Site personnel in accordance with the Health and Safety Plan (BBL, July 1998).

Ambient air monitoring was conducted at the Site to ensure protection of the surrounding community and the environment during operations. The ambient air monitoring was performed at the beginning of the sheet pile installation and at the beginning of the soil pressure grouting. Poly-Urethane Foam (PUF) and Total Suspended Particulate (TSP) systems were used to collect 24-hour ambient air samples at three locations (one upwind, two downwind) for the two sampling episodes completed during construction activities. The samples at each location, including the field blank, were analyzed for organics, metals and PCBs. Analytical results did not detect any constituent above applicable Federal and State Ambient Air Quality Standards.

Sheet Piling. The sheet piling quality assurance relied on the American Society for Testing and Materials (ASTM) Standards A 6, A 572, and A 709. The sheet piling installation was monitored for plumbness and alignment, and the presence of a water-tight joint sealing and the use of H-Piles in the event of encountering an obstruction.

The original installation plan was to drive the sheet pile to an elevation of 170.68 m(560.0 feet) above mean sea level (AMSL). However, after encountering several obstruction, H-Piles were first used to clear a path for the sheet piles through the landfill refuse to an elevation of 171.29 m(562.0 feet) AMSL. This was performed to prevent damage to the sheet piles, which could compromise the quality of the water-tight seal created between the sheet piles. Once a series of H-Piles were driven to depth of 171.29 m(562.0 feet) AMSL, they were removed and the sheet piles were driven to the original plan elevation.

The 30.5 cm(12 inch) wide H-Piles, which are much stronger then sheet piles, were installed utilizing a 90,703 kg(100 ton) crane with a pile driver equipped with a 3,628 kg(8,000 pound) hammer. The blows per 0.31 m (foot) penetration ranged from 4 to 29. The H-Piles were effective in clearing a path for the sheet pile installation.

PZ27 sheet piles from Nucor-Yamato Steel were driven in place following the removal of the H-Piles. The 45.7 cm(18-inch), "Z" shape sheet piles were installed first utilizing the 90,703 kg(100 ton) crane with the addition of a vibratory hammer. Each sheet was monitored during installation for plumbness and alignment which was necessary to ensure the proper installation of the sheet piles.

Adeka Ultraseal A-50 was applied to the joints of the sheet piles before installation to create a water-tight seal between the joints. The sealant cured to a solid rubber and was kept dry until the sheets were driven below the water table where the contact with the water cased the sealant expand to 4 times its original size. This action completed the seal process. The integrity of the sheet piling seal was checked by visual inspection throughout the installation of the sheet piles.

Groundwater Treatment for Grout Mixture. Based on results from the treatability study, a temporary water treatment plant was constructed on-Site to treat water pumped from dewatering the MW-8 Area during the soil pressure grouting. The treatment plant consisted of a 75,700 L(20,000 gallon) holding tank with an aerator for untreated water, a 37,850 L(10,000 gallon) settling tank for aerated water, followed by two 136 kg(300 pounds) carbon filters, and a 75,700 L(20,000 gallon) treated water tank. The treated water was utilized in the grout mix and injected back into the MW-8 Area to ensure that there was little or no contaminated water generated for treatment/disposal off Site during the soil pressure grouting process. A total of volume of approximately 77,100 L of water was treated and 66,500 L were used in the grout mixture. The remainder of the treated water, approximately 10,600 L and decontamination water, approximately 5,700 L was characterized and disposed of in accordance with applicable regulations.

Grout Mixing. The grout mix consisted of approximately 1,400 L(370 gallons) of water (treated or municipal), 341 kg(752 pounds) of cement, and 45 kg(100 pounds) of bentonite, per batch. The intent of the grout mix was to provide a grout with a viscosity that was close to water formation penetration would be maximized during grouting in order to fill the effective porosity of the subsurface materials. Density testing was performed every batch 3,785 L(1,000 gallons) and ranged from 1,265 to 1,410 kg per m^3(79 to 88 pounds per cubic foot). The viscosity testing was performed every 7,570 L of grout mixed using a Marsh funnel and ranged from 38 to 46 seconds. These values are close to the viscosity of water, which is 28 seconds.

Grout Field Test Program. Prior to completion of the full-scale grouting, the Field Test Program was completed to determine optimal spacing between grout borings and to refine other parameters. Field test parameters included items such as grout pressure, lifting interval, pumping rate, grout borehole spacing, and grout borehole opening distances (vertical opening length). Grouting was considered to be completed at each borehole when the maximum grouting pressure was reached.[1]

The field test program was conducted in a 12- by 12-m grid located at the southwest corner of the site. Three observation holes (O-1 through O-3) were installed in between the grout borehole locations to observe the lateral flow of grout during the test program.

Grout Borehole Installation. The grout boreholes were installed using an air rotary drill by either pushing or pneumatic hammering the 5.08 cm or 3.18 cm outside diameter (OD) pipe to the desired depth. In many areas, landfill obstructions were encountered, so many of the holes were predrilled utilizing a button bit with an air rotary drill. Initially, 20 primary boreholes and 26 secondary boreholes were installed, and an additional 38 tertiary boreholes were installed later.

Each grout borehole was completed with a series of 1.52-m(5-foot) lifts starting from the lowermost interval (A), which started within the very top of the Unit 3 clay. Once grout injection reached the maximum grout pressure, pumping was stopped temporarily while the grout pipe was raised 1.52-m to the next upper interval (B). Pumping was then resumed until the maximum grout pressure was reached. This approach proceeded until the entire Unit 2 interval was grouted (lift E being the uppermost 1.52-m interval). Table 1 summarized the amount of grout injected at each of the boreholes for each lift.

---

[1] Maximum grouting pressure is 3.5KN/m^2 per 0.31-m bls for unconsolidated soil. Use of greater pressures would result in lifting of the soil and creation of new voids rather than filling of existing voids. For a grout borehole completed to 9.14-m bls, the maximum grouting pressure was 103.4KPa.

TABLE 1. GROUT INJECTED AT PRIMARY, SECONDARY, AND TERTIARY BOREHOLES.

Borehole Type/Number	Grout Injected (L)
Primary/20	122,980
Secondary/26	67,770
Tertiary/38	120,020
Total	310,770

Grout Injection. The grout was injected at 3.5 KN/m^2 per 0.31 m bls at flow rates between 0 and 285 L per minute. The flow rate and volume of grout injected was monitored and recorded using a Brooks CRE Series 3560 magnetic flowmeter.

The primary and secondary grout borehole injections were completed simultaneously, followed by the tertiary grout borehole injections. Tertiary boreholes were installed until the area no longer accepted grout.

Grout injection began with the field test program, followed by grouting the primary and secondary holes adjacent to the sheet piling wall, and progressed inward towards the center with the subsequent primary and secondary holes. The purpose of this methodology was to force as much DNAPL and contaminated water to the center of the MW-8 Area in order to extract or ensure stabilization of the greatest amount of contaminants. The tertiary boreholes were injected with grout after the surrounding boreholes (primary, secondary) had been grouted.

Observation during the grouting process indicated that the grout was being installed as planned. Of the 37 tertiary boreholes installed, the final 20 boreholes accepted no or very little grout, which demonstrates that the effective porosity within the MW-8 Area was grouted and plugged to the maximum extent possible[2]. In addition, grout and in some cased trace amounts of DNAPL were observed in observation wells, monitoring wells, and completed grout boreholes. The water level within the MW-8 Area rose during grouting operations. Table 2 presents the grout borehole information.

Cap Installation. An impermeable cap was placed over the MW-8 Area that consisted of 40-cm layer of clay, compacted to 95% of the maximum dry density, followed by a 40-mil Flexible Membrane Liner (FML). An 60-cm layer of clay,

---

[2] "No grout take" is defined as exceedance of the maximum grouting pressure without pumping more than 11.4 L (3 gal) within the first minute after initiation of grout pumping for each of the three lifts within the borehole. The 11.4 L is used as the approximate volume necessary to fill the grout pipe itself.

compacted to 95% of the maximum dry density, was placed over the FML, followed by 20 cm of topsoil and vegetative cover. The cap was designed to match the expected final slopes for the Landfill. Figure 2 provides a cross-section of the capping installation.

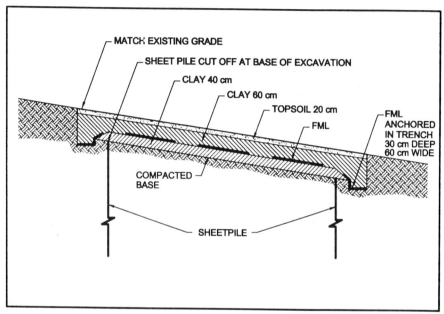

FIGURE 2. MW-8 Area capping details.

RESULTS AND DISCUSSION

Soil Test Borings. Soil borings were conducted so material could be recovered and inspected for grout, as requested by the MDEQ. The total volume of grout injected in the MW-8 Area represents approximately 12% of the total volume of Unit 2, which equates to a 12% effective porosity. Therefore, the grout recovered in the test borings was expected to be, on average, 12% of the total sample, and the spoons, with recovered samples generally exhibited this amount of grout. It is likely that some grout was squeezed out of the samples due to the compression caused by the hammering of the split spoon. Some split spoons exhibited little or no recovery, and others exhibited landfill material (e.g., glass, rubber pieces, and wood) that does not have effective porosity and therefore would not be expected to accept grout.

Grout Mold Testing. Six "wet grout" samples were cast in 7.62-cm by 15.24 cm cylinder molds and retained for compression strength testing. Three of the molds were created using the treated water grout mixture and three molds were created using the city water grout mixture. The results of the compression tests indicated that compression strengths ranged from 1.9 to 3.0 MPa (270 to 440 psi) for the treated water samples, and 2.96 to 3.9 MPa (430 to 560 psi) for the city water samples.

Operation and Management.   Operation and maintenance (O&M) activities consist of inspecting the cap for integrity and monitoring groundwater to ensure the continued effectiveness of the IRA components.   In addition, monitoring of hydraulic conditions will demonstrate that the groundwater is moving around the MW-8 Area and confirm that there is no hydraulic connection between the MW-8 Area and the surrounding water bearing unit.   Currently O&M is ongoing.

The completion capping of the MW-8 Area consisted of cutting off the top of the sheet pilings and leveling the surface to 120 cm below the original grade. A 40 cm compacted clay layer followed by a 40 mil flexible membrane liner (FML), 60 cm of compacted clay, 20 cm of topsoil to match the existing grade. The MW-8 Area IRA activities, as implemented, were completed for a total cost of approximately $1.4 million and are considered to be an effective remedy.

REFERENCES
Blasland, Bouck & Lee (BBL). June 1997. *Final Health and Safety Plan for Remedial Investigation/Feasibility Study Activities.* Syracuse, NY.
Blasland, Bouck & Lee (BBL). June 1998. *Interim Response Action Scope of Work.* Syracuse, NY.
Conestoga-Rovers & Associates, Inc. (CRA). January 1999. *Construction Completion Report, Interim Response Actions for the Monitoring Well 8 Area.* Romulus, Michigan.

CHARACTERIZATION AND REMEDIATION OF A MAJOR DNAPL SITE

Eric C. Lindhult, P.E. (Dames & Moore, Willow Grove, PA)
Michael J. Edelman, P.G and Thomas R. Buggey, P.G. (Dames & Moore, PA)
Michael A. Hart, P.E. (Carpenter Technology Corporation, Reading, PA)

ABSTRACT: During an environmental evaluation at a former manufacturing site, free-phase trichloroethylene (TCE) was detected during monitoring well installation. Subsequent careful evaluation delineated a pool of TCE, a dense non-aqueous phase liquid (DNAPL), which had accumulated above a dense silty clay layer in the shallow overburden aquifer. A second pool of TCE was subsequently detected beneath the floor of the manufacturing building. Active source and residual TCE remediation was performed through a multi-phase treatment train to remove contaminant mass and maintain hydraulic control. The treatment system has been in operation for 5 years and recovered several thousand gallons of free-phase and reconstituted TCE at an average rate of 5 to 10 gallons per day (gpd [19 to 38 lpd]).

INTRODUCTION

Site History – The 40-year-old former manufacturing facility is located in an urban area adjacent to other commercial properties. Two potential entry zones for free-phase TCE have been identified: a transfer pipe from a former aboveground storage tank (AST) and a former vapor degreasing unit.

A bedrock production well has operated on the site since the early 1950s, continuously pumping up to 200 gallons per minute (gpm [750 lpm]). The axis of the cone of depression from pumping at 200 feet (61 m) below ground surface is oriented east to west (Figure 1). This is not consistent with the strike and dip of local bedrock units suggesting that primary groundwater flow direction may be controlled by local fracture orientation.

Geologic Setting - Bedrock in the area consists of interbedded red shale, siltstone, and thin to thick-bedded sandstone, and strikes northeast with dips from 5 to 12 degrees northwest. Concentrations of TCE decrease dramatically with depth between the two upper bedrock zones that are monitored, though there is no obvious lithological or hydrogeological evidence of a lower permeability unit (aquitard) between the two zones that would inhibit vertical migration of TCE in the groundwater. However, head measurements made during straddle-packer testing of the bedrock production well suggests a vertical gradient reversal from downward in the shallow zones to upward near the bottom of the well.

Overburden soils at the site consist of interbedded silt, sand, and cobbles in a silt matrix. There are two prominent hydrostratigraphic overburden units. The upper overburden is a silty sand unit that grades laterally to a clean coarse sand. The lower overburden is a glacial till consisting of dense silt, clay, and some

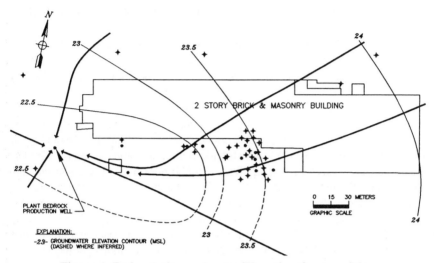

Figure 1. Bedrock Groundwater Elevation Contour Map

gravel of varying proportions. The two overburden units appear to be separated by a dense silt layer upon which a DNAPL pool has been observed. This layer contains variable amounts of clay with a maximum clay content of 20 percent. Groundwater is under unconfined to semi-confined conditions in the deep overburden zone. The TCE pools are within the saturated overburden.

INVESTIGATION AND MONITORING
Groundwater Monitoring of TCE Concentrations and Product - Monitoring wells were installed at the site in phases starting in 1984. Wells installed during the 1980s focussed upon evaluating concentrations associated with an entry zone near a former AST. A leaking pipe associated with the AST is suspected of creating the entry zone. Recovery of free-phase product in this area removed nearly all of the mobile-phase TCE above the dense silt layer and stabilization of the residual phase (apparent through the absence of mobile-phase in an area that once supported product up to 3-feet (1-m) thick in some of the observation wells). Groundwater flow in the shallow overburden is to the southwest, south, and southeast in a radial pattern. No significant gradient changes have occurred in this area since establishing a zone of capture. Mobile-phase TCE was recovered concurrent with monitoring of horizontal and vertical gradients to prevent changing gradients and exceeding the entry pressure of the DNAPL through the silt.

An additional entry zone near a former degreaser was identified, investigated, and confirmed. The assessment of former degreasers was achieved via soil sampling and screening using a hydrophobic dye-testing kit available on the market. The confirmed degreaser entry zone was investigated using a soil vapor survey, soil borings and monitoring wells with a 15-foot (4.5-m) spacing. All monitoring wells at the site have been designed with a small sump at the bottom that extends partially into, but not penetrating the dense silt. Dye testing of soil borings suggests non-continuous layers of product-bearing soil through the saturated soil column. Hence the sump at the base represents a collection point for the mobile-phase product.

No deep overburden or bedrock wells were installed in the immediate vicinity of the DNAPL pools as a precaution against creating a vertical migration pathway. Deep overburden wells are situated upgradient and downgradient of the shallow overburden entry zones to monitor if significant changes in the deep overburden hydraulic and chemical gradients have occurred. No mobile-phase TCE has been observed in deep overburden or bedrock wells to date. Elevated concentrations of TCE, on the order of 20 mg/l, have been observed in deep overburden and bedrock wells. These concentrations appear to increase toward the potential entry zones in the shallow overburden. Historic pumping of the plant production well appears to have achieved a capture of groundwater while preventing off-site migration of mobile-phase TCE. The containment of mobile-phase on site is inferred in the deeper zones from significant decreases in groundwater concentrations in off-site wells.

Site-specific Evaluation of the 1% of Solubility Rule of Thumb - Early in the history of DNAPL sites, USEPA adopted a multiple-point methodology to predict proximity to a DNAPL source at Superfund sites. Estimating the potential for DNAPL included a 1% of solubility rule of thumb in conjunction with other determinations including soil concentrations and dye testing. The 1% of solubility rule of thumb is a useful tool to assess the potential for the presence of DNAPL upgradient of a particular dissolved plume. The State agency involved with this site has placed into their guidance the 1% of solubility indicator as representing DNAPL presence at a specific well location. This is an application inconsistent with the original purpose of establishing criteria for predicting the potential for DNAPL. The State has indicated the intent to use the 1% of solubility as an indicator of immediate proximity to DNAPL unless a scientifically defensible site-specific alternative is demonstrated.

Because there are concentrations of TCE at 1% of solubility (11 mg/l) downgradient of the known source areas at the site, the rule of thumb appears to work as intended by the USEPA. However, the use of the rule of thumb as a broad-based indicator of immediate proximity to a DNAPL is not consistent with site data. A world-recognized technical expert was contracted to prepare a site model to assess the potential distance that a concentration of 1% of solubility can be observed from a DNAPL source. For the overburden, Monte Carlo simulations indicated that dissolved concentrations on the order of 10% of solubility of TCE could be observed at a distance of hundreds of feet downgradient of a DNAPL source. For the bedrock, the modeling suggested that the 1% concentration could propagate for distances greater than 1,000 feet (300 m) without indicating immediate proximity to DNAPL, due to specific aperture spacing for fractures in bedrock beneath the site (a wide range of spacing was used for the model to asses the model sensitivity to this variable). Contrary to the 1% of solubility rule of thumb, some on-site wells contain dissolved TCE up to 500 mg/l (45% of solubility) in the absence of any indication of DNAPL present.

Natural Attenuation Indicators - Preliminary review of natural attenuation parameters, primarily chloride and sulfate, suggest natural attenuation may be occurring at the property boundary, approximately 150 feet (45 m) from the second entry point. The trend analysis for chloride and sulfate is complicated by relatively elevated background concentrations. Although chloride concentrations appear to decrease toward the property boundary, the data is not sufficient to define an obvious trend. Sulfate poses a similar issue in that variations across the site do not provide an obvious trend. Further monitoring and evaluation of daughter products of TCE degradation including dissolved gases (methane, ethane, vinyl chloride, etc.) and field parameters (DO, ORP, EC, etc.) is planned to evaluate natural attenuation.

SITE REMEDIATION
Remediation System Design - A preliminary on-site remediation program was designed to recover the mobile-phase TCE (mass removal); control migration of volatile organic compound (VOC) contamination in both aquifer systems; remove VOCs from the vadose zone; and remove residual VOCs from treated air and water streams prior to discharge to the atmosphere or storm sewer. Among the design considerations was potential demolition of the existing facility to allow for future site development. The design was submitted to the state regulatory agency and approved for on-site remediation. Key system components are shown on Figure 2.

Groundwater and Soil Remediation System (GRS) Description - Approximately 50 gpm (190 lpm) of mobile-phase TCE and contaminated groundwater is recovered by pumps from 13 overburden recovery wells. The shallow overburden wells are connected to a VE blower that helps remediate the soil and enhances the recovery of groundwater and mobile-phase TCE from the wells. Bedrock groundwater is recovered from one well.

Two countercurrent air stripping towers remove the majority of the VOCs from the recovered groundwater. One tower is used for the overburden flow after DNAPL separation and pretreatment, and the other tower is used for the bedrock flow (75 gpm / 285 lpm), which does not receive any pretreatment. The combined groundwater flow from these towers is then polished by two GAC units, prior to discharge to the storm sewer.

The combined offgas from the air stripping towers, VE blower, and aeration tank is preheated to reduce the relative humidity and is then passed through two vapor-phase GAC units in series. Several times per day the lead GAC unit is removed from service and injected with steam to desorb the accumulated VOCs, primarily TCE. The steam and vapor mixture is then cooled

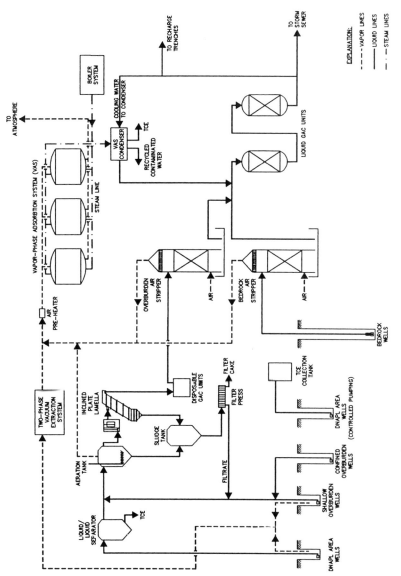

Figure 2. Groundwater and Soil Remediation System

in a condenser, and the TCE is recovered for off-site reprocessing. A computer-based system is used to monitor over 100 input and output signals from the various instruments and motors, and can vary operational conditions or shut down the system if operating conditions are outside acceptable ranges. This allows for continuous system operation with minimal operator attendence.

GRS Operation - The full-scale GRS was initiated in July 1994 (before the second TCE source was discovered in 1995), after final approval of operating permits. Prior to startup, the groundwater elevations indicated flow in the shallow overburden was radial to the southeast and southwest from the center of the site, and in a westerly direction in the deep overburden (40 to 80 feet [12 to 24 m] deep) and bedrock (80 to 200 feet [24 to 60 m] deep) zones. Little to no DNAPL was observed in the wells using clear bailers, even in wells in the known free-phase area.

Startup - During the first several months, the overburden groundwater flow was 40 to 55 gpm (150 to 210 lpm), which was as expected. The flow from the bedrock was approximately 90 gpm (340 lpm), and the VE was operated at approximately 7 to 10 in. mercury (Hg [0.23 to 0.33 atmospheres]). During the initial months, several hundred gallons of DNAPL were recovered from the mobile-phase area by a liquid/liquid separator. The groundwater elevations near the recovery wells initially were significantly lower due to pumping. After several months, the water table stabilized at the same general time that the overburden groundwater recovery rates decreased to 20 to 30 gpm (75 to 115 lpm).

Recovery of reconstituted TCE from the vapor-phase GAC units was approximately 2 to 4 gpd (7.5 to 15 lpd) during the startup operation (part-time operation in the overburden). The recovery increased noticeably to approximately 6 to 12 gpd (22.5 to 45 lpd) when the system started operating 24-hours per day in December 1994. The recovery decreased slightly over time as the overburden groundwater recovery rates continued to slowly decrease and as the VE system removed TCE from the vadose zone. By July 1995, the TCE recovery rate was generally less than 5 gpd (19 lpd).

Overburden System Modification - After the GRS startup period, steps were taken to optimize system performance. The objective to maintain hydraulic control required water table depression. Depression of the water table in the overburden was also desired to expose additional soil to the VE system to enhance mass removal. To meet these objectives, new strategies were initiated to enhance groundwater recovery and mass removal. The primary initiatives were 1) increasing the vacuum to remove groundwater through the dual-phase VE system (simultaneous recovery of groundwater and soil vapors), 2) applying a vacuum to wells installed inside the building, 3) installing electric pumps, and 4) maximizing dual-phase VE.

Increase to Vacuum - After several months of operation, the vacuum to the system was increased to 15 in. Hg (0.5 atmospheres). At this vacuum, groundwater was recovered at approximately 14 gpm (50 lpm). This program was initiated in September 1995. TCE recovery in the vapor-phase GAC units increased to approximately 8 gpd (30 lpd). A slight, but noticeable drop in water levels in the shallow overburden was observed.

Vacuum on Interior Wells - With the increased vacuum on the system, we also evaluated methods to increase mass removal, including applying a vacuum on the wells in the vadose (unsaturated) zone underneath the building. The two wells were installed in the bottom of an earthen trench during the building decommissioning activities in 1988. Although hydrophobic dye tests of this trench did not detect a potential DNAPL entry zone, photoionization detector readings indicated that TCE might have been discharged at this location.

After applying full vacuum to the interior wells, TCE recovery from the VAS increased dramatically to over 15 gpd (55 lpd) for 21 days. The recovery then decreased to approximately 10 gpd (38 lpd), though still greater than during the single-phase VE operation.

Installation of Electric Pumps - Due to the persistent problems with the silting of the pneumatic overburden pumps at the site, several electric pumps were evaluated as replacements. Several submersible pumps and two centrifugal jet pumps were installed. The submersible pumps appeared to perform better than the jet pumps, whose performance is impacted by the depth to water and the need to overcome the applied vacuum in the shallow overburden.

Groundwater recovery noticeably increased with the electric pumps. No problems were observed with the electric pumps due to silt problems. The electric pumps resulted in fewer shutdowns and improved the efficiency of the GRS. The water levels lowered an additional 3 feet (1 m).

Optimizing Two-Phase Vacuum Extraction - Several options were evaluated to further lower the water table and expose more soil to VE. The removal of VOCs from underneath the building would remove possible source areas and reduce the total TCE mass at the site. The ultimate goal was to reduce the treatment time. After evaluating options, a high-vacuum liquid-ring vacuum pump (LRVP) was selected to increase the vacuum and flow rate to the GRS. The LRVP operation was initiated in July 1996 and overburden groundwater recovery rates increased to approximately 22 gpm (85 lpm). The vacuum increased to 20 in. Hg. (0.67 atmospheres), resulting in decreasing water levels and increased TCE recovery.

New Source Area - Low-volume pumping using pneumatic bladder pumps began in May 1996 to recover mobile-phase TCE. Bladder pumps with individual controllers were installed in nine wells. After 7 months, over 1,000 gallons (3,785 liters) of free-phase TCE had been recovered. The bladder pumps were used for low-volume pumping to induce TCE flow toward the wells (referred to as passive water flooding). Observations indicated that the TCE continued to flow towards the well during low-volume pumping. However, TCE required longer periods to reappear in a well if the pool was pumped too quickly (i.e., greater than the recharge rate into the well).

Options are continuously identified and evaluated to enhance mass removal in the area. Options being considered include:

· Sheet piling to isolate the area
· Potassium permanganate addition to oxidize the TCE
· Aggressive pumping to reduce the hydrostatic head on the free-phase TCE to enhance the flow of product to the well for recovery
· Apply dual-phase VE to aggressively dewater the area, expose additional soil to VE, and greatly reduce the hydrostatic head on the TCE to enhance the flow of product to the well for recovery

Lessons Learned - Although the investigation, remedial design, and GRS operation have proceeded without major problems, several lessons were learned from this project concerning DNAPL TCE sites and pump-and-treat systems in general:

· Where possible, install a sump on all wells suspected to be in a DNAPL area. The sump, should be placed only a few inches into the top of a suspected confining layer, so that the entire thickness of the DNAPL can drain into the well. However, care must be taken in designing the well and selecting material so as not to promote DNAPL vertical migration (loss of pool). Bentonite should be installed in the well bottom and around the sump so as to minimize the potential conduit affect for any DNAPL in the well's sand pack.
· The use of dual-phase VE, which removes VOCs in the soil matrix, enhances groundwater removal and possibly increases TCE flow to the recovery well, appears to be promising as an effective remediation technique for DNAPL in overburden aquifers.
· DNAPL in the fine-grained silt presented a problem due to silt entrapment during total fluids pumping. Pneumatic pumps may not have been the best selection for this application due to silting. Electrical pumps were the better solution for wells with reasonable yield and construction.
· The flexibility built into the system is critical, because the final remediation program is likely to change, such as adding pumping wells, re-piping of flow to different equipment, greater flow rate capacity, etc. DNAPL remediation should be approached gradually and in steps, because DNAPL does not behave like groundwater. Modification to the system should occur incrementally, observing the results and adjusting the system to achieve the desired result (mass reduction, lowering the water table), without losing the DNAPL pool.

A POLYMER FLOOD TO RECOVER DNAPL FROM SHALLOW FRACTURED BEDROCK

Rawson, James R. Y. (GE - Corporate R&D, Niskayuna, NY), Bridge, Jonathan R., Guswa, John H. (HSI GeoTrans, Inc., Harvard, MA) and LaPoint, Edward (GE – Corporate Environmental Programs, Albany, NY)

ABSTRACT: A small-scale pilot polymer flood was carried out in a shallow fractured bedrock zone to facilitate the recovery of DNAPL containing PCBs. The fractured bedrock zone was located on the dry face of a waterfall, immediately adjacent to an old manufacturing plant that had assembled capacitors containing PCBs. For a number of years, DNAPL containing PCBs seeped from the fractures in this zone and was manually recovered from low spots on the dry face of the waterfall.

The purpose of the polymer flood was four fold. It was intended to: (1) accelerate and improve the recovery of DNAPL from certain fractures in the shallow bedrock; (2) reduce the potential for future DNAPL seepage from these fractures; (3) determine whether there was a localized DNAPL reservoir in these fractures; and, (4) evaluate whether a polymer flood could be used to recover DNAPL from other areas at the same site.

Prior to the polymer flood, water was injected into the shallow fractured bedrock. This waterflood was designed to determine the rate at which fluids would flow from the fractured bedrock and evaluate the effectiveness of the collection system that was to capture the polymer and DNAPL. The collection system was located immediately adjacent to the fractures from which DNAPL had been seeping onto the face of the dry waterfall. After the waterflood, a polymer was injected into the fractured bedrock. Both the polymer and the DNAPL were recovered in the collection system. Then a second waterflood was carried out in the same area to determine whether any more DNAPL would seep from the shallow fractured bedrock during normal groundwater discharge.

The polymer flood recovered approximately four times more DNAPL than did the first waterflood. Comparison of the composition of the PCBs in the DNAPL recovered during the polymer flood to the PCBs in the DNAPL recovered during the two waterfloods showed that the polymer flood contacted DNAPL that had not previously been swept by water.

INTRODUCTION

During the fabrication of electrical capacitors at an old manufacturing plant, there were a number of incidental releases of polychlorinated biphenyls (Aroclor 1242) to the environment. Although the use of polychlorinated biphenyls (PCBs) was discontinued in 1977, PCBs have since migrated as DNAPL from the basement of the main manufacturing building at the plant site into fractured bedrock and subsequently to the face of what is today normally a dry waterfall.

The geology immediately beneath this old plant site consists of a layer of unconsolidated deposits, up to 21 feet thick. The unconsolidated deposits are composed of glaciofluvial outwash, lucustrine clay, till and artificial fill. Beneath the unconsolidated deposits is fractured shale, which ranges in thickness from 150 feet to 259 feet. The shale overlies two distinct limestone formations, which together are approximately 150 feet thick. A portion of the foundation of the main manufacturing building at the site was excavated into the shale bedrock. Several drainage structures, tunnels, sewers and air plenums were also excavated into the bedrock beneath the floor of this building. The primary pathways of DNAPL migration into the shale have been through vertical and horizontal fractures. The fractured shale is exposed on cliffs and a dry waterfall immediately adjacent to the site. In addition, two major parallel sub-horizontal thrust fault planes, the upper and the lower fault planes, are exposed on the dry waterfall and provide major pathways for the migration of DNAPL.

A dam is located at the top of the waterfall. The dam diverts the flow of a large river through a hydroelectric plant on the opposite side of the river from the old plant site. The total drop from the top of the dam to the base of the falls is 67 feet. The waterfall itself drops approximately 40 feet. During much of the year, the total river flow (3000 to 5000 cfs) is diverted through the hydroelectric plant and the waterfall is dry. Periodically, during routine maintenance at the hydroelectric plant and during high river flow (greater than 8000 cfs), water will spill over the dam onto the waterfall.

Subsequent to diverting the river through the hydroelectric plant and prior to initiating remedial measures at the site, DNAPL seeped from the shallow fractured shale onto the dry face of the waterfall. In one area, DNAPL seeped from fractures associated with the upper fault plane in the shale into a low area on the dry waterfall. Over a period of several years, approximately 1.4 liters of DNAPL was manually collected from this low spot on the waterfall. More recently, an enclosed drain was constructed to facilitate collection of the DNAPL that seeped from the upper fault plane in this area. This collection system precluded any discharge of DNAPL to the river during high flow conditions.

A polymer flood test was carried out in the aforementioned area of the shallow fractured bedrock to: (1) accelerate and improve the recovery of DNAPL; (2) reduce the potential for future DNAPL seepage; (3) determine whether there was a localized DNAPL reservoir in the shallow bedrock; and, (4) evaluate whether a polymer flood could be used to recover DNAPL from other selected locations at the site.

CONCEPTUAL DESIGN OF POLYMER FLOOD

A pressure differential can be used to cause mobile oil or NAPL in a permeable geologic media to migrate to a recovery system. The effectiveness of a pressure differential to cause this to occur is dependent upon the mobility ratio (M) of the fluids (water and NAPL) in the permeable geologic media (Chang, 1978; van Poollen, 1980; Lake, 1989).

The mobility ratio (M) of fluids (water and NAPL) in a permeable geologic media is defined in equation 1.

$$M = \frac{\text{mobility of } H_2O}{\text{mobility of NAPL}} \qquad (1)$$

In equation 1, the mobility of water is equal to the permeability of the geologic media to water (H_2O permeability) divided by the viscosity of the water (H_2O viscosity). In this same equation, the mobility of NAPL is equal to the permeability of the geologic media to the NAPL (NAPL permeability) divided by the viscosity of the NAPL (NAPL viscosity). The water viscosity and the NAPL viscosity are the viscosities of water and NAPL, respectively, in the permeable geologic media. Hence, equation 1 can be expressed as shown in equation 2.

$$M = \frac{H_2O \text{ permeability} / H_2O \text{ viscosity}}{NAPL \text{ permeability} / N\tilde{A}PL \text{ viscosity}} \qquad (2)$$

If the viscosity of the water is increased by the addition of a viscous polymer, the mobility ratio of the water and DNAPL will be substantially altered. If for simplicity and expediency in discussion, one assumes that the permeability of the water is equivalent to the permeability of the DNAPL, then the mobility ratio (M) can be approximated as shown in equation 3.

$$M \simeq \frac{\text{viscosity of DNAPL}}{\text{viscosity of } H_2O} \qquad (3)$$

Simplification of equation 2 to equation 3 can be justified on the basis that the relative difference in viscosity between an aqueous polymeric solution (hereafter referred to as a polymer) and water is far greater than the relative difference in permeability of the water and the DNAPL in the geologic media.

Figure 1 shows a schematic that compares and contrasts the flow of fluids in a permeable geologic media during a waterflood and a polymer flood. In the case of a waterflood (Figure 1a), the mobility of the fluids (water and DNAPL) is greater than 1.0 (M > 1.0) and the water will be more mobile than the DNAPL. This results in water fingering around and past the DNAPL and only minimal flow of DNAPL.

In the case of a polymer flood (Figure 1b), the mobility ratio of the fluids (viscous polymer and DNAPL) in the permeable geologic media will be less than 1.0 (M < 1.0). This prevents fingering of the viscous polymer around the DNAPL and favors displacement of the DNAPL by the polymer. The result is an increased flow of DNAPL.

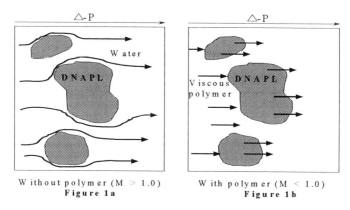

Without polymer (M > 1.0)
Figure 1a

With polymer (M < 1.0)
Figure 1b

FIGURE 1: Schematic showing the flow of fluids during a waterflood and a polymer flood. Figure 1a shows the flow of water and DNAPL during a waterflood. Figure 1b shows the flow of a viscous polymer and DNAPL during a polymer flood.

The viscosity of the water in the shallow fractured bedrock flood at the aforementioned site can be increased from 1.0 centipoise to several hundred centipoise by the addition of a variety of different types of polymers. Xanthan gum was used in this pilot test to increase the viscosity of the water. Xanthan gum will not chemically interact with the DNAPL in the fractured bedrock at the site and the viscosity of the DNAPL will remain constant (approximately 25 centipoise for Aroclor 1242).

IMPLEMENTATION AND ANALYSIS OF POLYMER FLOOD

Figure 2 shows the area on the waterfall where the polymer flood was carried out. This same figure also shows the location of the five injection wells used to inject water and polymer into the shallow fractured bedrock and the design of the collection system. These injection wells and the portion of the collection system used for the waterfloods and the polymer flood were located in a total area of approximately 210 sq. ft.

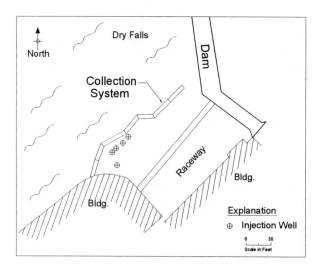

FIGURE 2: Schematic of area on the dry waterfall where the polymer flood was carried out.

Initially, a waterflood was used to flush DNAPL from the shallow fractured bedrock into the collection system. This waterflood consisted of simultaneously pumping 200 gallons of potable water at a rate of 6.7 gpm into the five injection wells shown in Figure 2. The resulting pumping rate maintained a head in these wells of approximately 2.7 to 3.9 psi. Within minutes after starting to pump water into the wells, water flowed from discrete zones of the upper fault plane into the capture system. Additionally, small droplets of DNAPL were seen to flow with the water into the collection system. The duration of the waterflood was approximately one half-hour. After the waterflood was completed, the water and the DNAPL in the collection system were pumped to a 220-gallon tank and mixed vigorously by continuously recirculating the fluids. Then three 1 liter samples of the recovered fluids (water and DNAPL) were collected for quantitative analysis of PCBs.

Next, a polymer flood was carried out in the same area. The polymer flood consisted of simultaneously pumping 200 gallons of xanthan gum (1500 ppm dissolved in potable water) at a rate of 3.7 gpm into the same five injection wells used in the waterflood. The resulting pumping rate maintained a head in these wells of approximately 2.9 to 3.8 psi. Within minutes after starting to pump the polymer into the wells, fluids flowed from discrete zones of the upper fault plane into the collection system. Visual observations indicated that more DNAPL flowed into the collection system during the polymer flood than during the waterflood. The entire polymer flood was completed in approximately one hour. The polymer and the DNAPL in the collection system were recovered, mixed and sampled for PCB analysis as described for the first waterflood.

A second waterflood was used to flush DNAPL from the same area into the collection system. This waterflood consisted of simultaneously pumping 200 gallons of potable water at a rate of 4.3 gpm into the same injection wells. The resulting pumping rate maintained a head in these wells of approximately 2.8 to 4.6 psi. Within minutes after starting to pump water into the wells, water flowed from discrete zones of the upper fault plane into the collection system. The second waterflood took approximately 0.85 hours. Again, the water and the DNAPL in the collection system were recovered, mixed and sampled for PCB analysis as described for the first waterflood.

More than 95% of the fluids injected into the wells during the two waterfloods and the polymer flood were recovered in the collection system. In addition, no fluids were observed exiting the shallow fractured bedrock in any location other than from the upper fault plane into the collection system.

Samples of total fluids from the two waterfloods and the polymer flood were analyzed for PCBs using EPA Method 8082. This analytical method uses a gas chromatograph (GC) with an electron capture detector (ECD) to quantify PCBs relative to a standard (Aroclor 1242). To assure complete extraction of the PCBs from the polymer, the xanthan gum was hydrolyzed with 0.25% sodium hypochlorite. For consistency in analysis of samples and interpretation of data, the samples collected from the waterfloods were similarly treated with sodium hypochlorite.

Table 1 shows the results of the quantitative analysis of the PCBs in the fluids collected from both waterfloods and the polymer flood.

TABLE 1: Quantification of PCBs recovered during waterfloods and polymer flood.

Fluid	Volume Fluids Recovered (gal)	PCB Concentration Ave, n = 3 (ug/L)	St Dev PCBs (ug/L)	Total Mass PCBs Recovered (gm)
1st Waterflood	200	2737	762	2.1
Polymer flood	200	10207	548	7.7
2nd Waterflood	200	3747	136	2.8

The data in Table 1 shows that the polymer flood recovered nearly four times more PCBs than the first waterflood. In addition, the recovery of PCBs in the second waterflood was approximately one third of that recovered during the polymer flood and was similar to that recovered during the first waterflood. This same data also shows that no significant mobile volume of DNAPL was recovered from the shallow bedrock by either of the waterfloods or the polymer flood.

The GC chromatograms of the PCBs recovered from the two waterfloods and the polymer flood are shown in Figure 3. This same figure also shows a GC chromatogram of an Aroclor 1242 PCB standard. The most notable feature seen in these chromatograms is the variation in the quantity of PCB peak 5 (PCB-5).

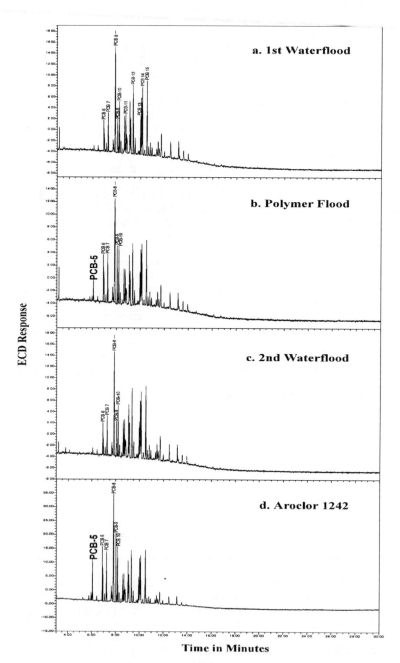

FIGURE 3: GC chromatograms of PCBs recovered from the waterfloods and the polymer flood. Extracts from the samples collected from the two waterfloods and the polymer flood were analyzed in a GC using an ECD and compared to a standard Aroclor 1242 sample.

When Aroclor 1242 is analyzed using EPA Method 8082 (Figure 3d), PCB-5 is a notable component of the PCBs. PCB-5 was nearly absent in the DNAPL recovered during both the first and second waterfloods (Figures 3a and 3c). In contrast, PCB-5 was present in the DNAPL recovered during the polymer flood (Figure 3b). This variation in the content of the PCBs in the DNAPL recovered during the two waterfloods and the polymer flood can be explained on the basis of the differential aqueous solubility of the two PCB isomers (congeners) in PCB-5 and the remaining PCBs in Aroclor 1242. The solubilities of the two congeners in PCB-5 are 560 and 880 μg/L. In contrast, the overall solubility of the aggregate of the PCBs in Aroclor 1242 is approximately 250 μg/L and ranges from less than 0.1 to 880 μg/L for the individual congeners in this Aroclor mix.

The near complete absence of PCB-5 in the DNAPL recovered during the first and second waterflood indicates that this DNAPL had already been extensively contacted and flushed by water during groundwater seepage. This suggests that the DNAPL recovered during the two waterfloods was in a path through which groundwater flowed in the shallow fractured bedrock. The increased quantity of PCB-5 in the DNAPL recovered during the polymer flood indicates that the polymer contacted DNAPL which had not been previously flushed by any large flow of groundwater.

This latter observation is the first demonstration ever that a polymer flood actually contacted and recovered NAPL that had not been previously influenced by the flow of water. Only the unique differential physical chemical properties of the individual congeners in the PCBs in the DNAPL at this site permitted us to distinguish the difference between DNAPL recovered by the waterfloods from DNAPL recovered by the polymer flood. In summary, the individual congeners in the PCBs in the DNAPL provided unique tracers that were used to increase our understanding of the multiphase fluid flow of the DNAPL and water in the fractured bedrock at the site.

Finally, the fluids (water, polymer and DNAPL) were disposed of in a permitted on-site water treatment plant. The capacity of the water treatment plant was 125 gpm and consisted of a treatment train that included coagulation, clarification, multi-media filtration, ultraviolet–chemical oxidation (UV-hydrogen peroxide), air stripping and granular activated carbon filtration. The UV-hydrogen peroxide treatment was designed to destroy organic contaminants, including PCBs, in the water. The fluids recovered from the polymer flood were held separately from other water sources for ten days to permit the natural biodegradation of the xanthan gum. During this time, the viscosity of the fluids decreased to near that of water. When the resulting fluids were introduced into the influent of the water treatment plant, there were no upsets in the process parameters used to monitor the treatment plant.

CONCLUSIONS

The results from this test indicated that: (1) the fractured shallow bedrock on the waterfalls was highly permeable through which water or a polymer could flow with ease; (2) there was no large volume (liters) of DNAPL in the shallow fractured bedrock in the area of the pilot test that could be recovered by either a waterflood or a polymer flood; (3) the polymer flood increased the recovery of DNAPL from the fractured shallow bedrock; and, (4) the polymer flood accessed and recovered DNAPL that had not been previously been influenced by the normal flow of groundwater.

AKNOWLEDGMENTS

The authors wish to thank Keith Rudman for his expert technical assistance in carrying out the two waterfloods and the polymer flood.

REFERENCES

Chang, H.L. 1978. Polymer Flooding Technology - Yesterday, Today and Tomorrow. *J. Petroleum Technology*, August: 1113-1128.

Lake, L.W. 1989. Polymer Methods. in *Enhanced Oil Recovery*. pp 314-353. Prentice Hall, Englewood Cliffs, NJ.

Van Poollen, H.K. 1980. Polymer Flooding. in *Fundamentals of Enhanced Oil Recovery*. pp 83-103. PennWell Publishing, Tulsa, OK.

DISSOLUTION OF A RESIDUAL DNAPL IN VARIABLE APERTURE FRACTURES

S.E. Anderson and N.R. Thomson
University of Waterloo, Waterloo, Ontario, Canada

Abstract: This work focuses on the mechanisms affecting the dissolution of residual DNAPLs in variable aperture fractures. A conceptual model for the dissolution of residual DNAPLs in fracture planes is developed. In addition, results from a series of laboratory-scale dissolution experiments in laboratory-induced horizontal fracture planes are presented and discussed. In general, it was found that the percentage of mass removed from fracture planes through dissolution was somewhere between typical percentage mass removals reported for residuals and pools in porous media.

INTRODUCTION

Over the past few decades, dense non-aqueous phase liquids (DNAPLs) have been identified as an important class of groundwater contaminants. DNAPL releases to the subsurface often occur over fractured geologic deposits, and DNAPLs can migrate through fracture networks under the influence of gravitational and viscous forces. Capillary forces resist the forward movement of a DNAPL mass, and are responsible for separating and immobilizing DNAPL ganglia from the DNAPL mass. When a DNAPL ganglion becomes immobile, the dissolution process will govern the length of time that the DNAPL will persist. In fractured porous media, dissolution can occur both within the fracture plane and into the porous matrix surrounding the fracture.

It has been shown that DNAPL disappearance from a fracture is only significantly affected by matrix diffusion in relatively porous environments (Parker et al., 1994; VanderKwaak and Sudicky, 1996). Esposito and Thomson (1999) developed a numerical model to simulate aqueous phase transport in a variable aperture fracture, and coupled it with an existing two-phase flow model (Murphy and Thomson, 1993). It was shown that the time frame for NAPL dissolution is markedly sensitive to the mass transfer coefficient. It was also found that the hydraulic gradient across the fracture plane can aid in the removal of pure phase mass; however, diffusion-controlled mass transfer is necessary to dissolve NAPL that is trapped in regions of the fracture plane which are not accessible to aqueous phase flow. Glass and Nicholl (1995) simulated a fracture plane using etched glass to study the dissolution of an entrapped air phase. They were not able to explain their results using any of the existing conceptual models for dissolution, and concluded that further work is necessary in order to understand entrapped phase dissolution in fractured media.

The purpose of this paper is to develop a conceptual model for the dissolution of a residual DNAPL in a variable aperture fracture, and to use this conceptual model to help explain the observations from a series of laboratory experiments.

THEORY

It is commonly assumed that a linear driving force expression is suitable to characterize the interphase mass transfer flux, which can be expressed as (Powers et al., 1994)

$$J = k_f (C_s - C)$$ (1)

where k_f is a mass transfer coefficient, C is the aqueous phase concentration, and C_s is the equilibrium aqueous phase concentration. In the porous media literature, several investigations have suggested that the mass transfer coefficient is dependent on the diffusivity of the DNAPL component, and the rate of groundwater flow around the DNAPL (e.g. Powers et al., 1994). The mass transfer rate per unit volume of residual DNAPL can be calculated by multiplying the mass transfer flux by the specific interfacial area between the phases per unit volume of residual DNAPL. Thus, the spatial distribution of a residual DNAPL will affect its mass transfer rate to the aqueous phase by influencing the interfacial area available for mass transfer. In porous media, NAPL entrapment and residual saturation are influenced by the geometry of the pore network, the fluid-fluid properties (interfacial tension, viscosity and density ratios), the fluid-solid properties (wettability), and external forces on the fluids (pressure gradients and gravity).

Chatzis et al., (1983) presented two conceptual models used to visualize the different mechanisms by which NAPL may become trapped in pores. In the first model, NAPL may only become discontinuous when the pore body to throat ratio is large enough such that water tries to push NAPL through the pore throat at a rate which is faster than the NAPL is able to move through the pore throat. This is referred to as the snap-off mechanism. The second conceptual model, referred to as the pore doublet model, involves a tube splitting off into two pores and rejoining again downstream. In this model the water chooses the least resistant path and travels around the NAPL-filled pore thereby trapping the NAPL.

To the knowledge of the authors, no work has been reported in the literature to date regarding the mechanisms controlling the spatial distribution of residual NAPL in fracture planes. However, the variable nature of asperities in fracture planes is analogous to pore throats and bodies in porous media. Therefore, the conceptual models that exist to explain trapping mechanisms in porous media are conceivably applicable to fractured media.

EXPERIMENTAL DESIGN

To develop a conceptual model for dissolution in fracture planes, it was necessary to gain an understanding of the residual entrapment geometries and interfacial areas that might be expected in fractured media. This was achieved by conducting a series of visualization experiments in a transparent cast of a fracture plane. Dissolution experiments were then performed, under a range of hydraulic gradients, in two laboratory-induced dolomite fracture planes. In addition, hydraulic and tracer studies were conducted on both the synthetic and the laboratory-induced fracture planes.

Visualization Experiments. A series of experiments was conducted in a transparent cast of a fracture plane to observe two-phase flow and entrapment in variable aperture fractures.

A transparent cast of a laboratory-induced rough-walled dolomite fracture plane was fabricated. The casting procedure consisted of creating a silicone negative of each fracture face (RTV 3032, Rhône Poulenc Silicones VSI). The silicone negatives were then used to cast transparent positives of each fracture face using a chemically-resistant epoxy (Stycast 1269A, W.R. Grace Co.). Both the silicone and the epoxy were vacuum degassed before they were cast to ensure that gas bubbles would not compromise the integrity of the artificial surface. Finally, the artificial fracture faces were mated to form the transparent fracture plane.

The experimental apparatus (Figure 1) consisted of a glass flow cell fixed to the upstream and downstream ends of the fracture cast. Stainless steel putty was used to seal the remaining two sides of the fracture cast to create no-flow boundaries. A peristaltic pump was employed to inject water through the artificial fracture plane at selected flow rates, and piezometers were attached to the upstream and downstream ends of the fracture cast to permit the measurement of the hydraulic gradient across the fracture cast.

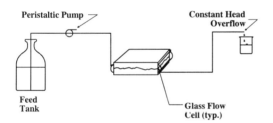

FIGURE 1. Schematic of experimental apparatus used for the dissolution experiments.

The visualization experiments were conducted by first saturating the fracture plane with water which was dyed red using food colorant. Once the saturation phase was complete, the DNAPL (HFE 7100, 3M) was trapped at a residual saturation in the horizontal fracture plane. HFE 7100 was chosen as the DNAPL because its physical properties are similar to those of many DNAPLs commonly found in the subsurface (see Table 1); however, it is non-toxic and it did not attack the epoxy cast. The entrapment process involved releasing the DNAPL into the fracture plane at a capillary pressure of approximately 3.5 KPa, which was sufficient to invade aperture regions larger than approximately 10 μm. Finally, water was flushed through the fracture plane at a capillary number (Ca) of 2.5E-3 in an attempt to mobilize as much of the DNAPL as possible.

Dissolution Experiments. A series of experiments was conducted to investigate the dissolution of single-component residual DNAPLs in horizontal variable-aperture fractures. Two laboratory-induced fracture planes were used in these experiments. The fractures were induced by means of applying a uniaxial load to dolomite samples containing stylolites. The stylolites, formed by the cementation of solution channels, acted as a plane of weakness and provided an ideal path for the tension fracture to follow under the uniaxial load.

The experimental apparatus was very similar to that used for the visualization experiments (Figure 1). The glass flow cells were well-mixed throughout the dissolution experiments. In addition, they were equipped with DNAPL exit ports to allow for the separation of any DNAPL from the aqueous phase.

The primary DNAPL used in these experiments was 1,1,1 trichloroethane (1,1,1 TCA). A subsequent experiment was conducted with trichloroethylene (TCE) for comparison purposes. Table 1 gives the relevant physical properties of these DNAPLs. Both compounds were dyed with Sudan IV (<8.5 mg/L) so that the presence and location of pure phase could be observed in the glass flow cells. Interfacial tension tests revealed that the presence of the dye did not significantly change the interfacial tension of either compound.

The dissolution experiments were conducted by first saturating the fracture plane with buffered water to prevent any dissolution of the dolomite fracture plane. After the fracture was water saturated, the DNAPL was trapped at a residual saturation in the horizontal fracture plane. The entrapment process involved releasing the DNAPL into the fracture plane at a capillary pressure of approximately 3.5 KPa, which is sufficient to invade aperture regions larger than approximately 20 μm and 15.5 μm for 1,1,1 TCA and TCE respectively. Water was then flushed through the fracture plane at a Ca of 2.5E-3 in an attempt to mobilize as much of the DNAPL as possible. The mass successfully mobilized was removed from the effluent flow cell by means of the DNAPL exit port. The mass entrapped was determined from mass balance considerations. Finally, water was flushed through the fracture plane under a range of flow rates (10.1 m/day to 104.2 m/day), and the effluent concentration profile was measured until steady-state conditions were observed.

TABLE 1. Physical properties of organic phases.

Property	Symbol	Units	1,1,1 TCA	TCE	HFE 7100
Density	ρ	$g \cdot mL^{-1}$	1.33§	1.46§	1.52§
Solubility	C_s	$mg \cdot L^{-1}$	1300*	1100*	< 20¶
Interfacial Tension	σ_{n-w}	$dynes \cdot cm^{-1}$	36.7¦	29¦	42.4¥
Viscosity	μ	$g \cdot cm^{-1} \cdot s^{-1}$	0.62*	0.39*	0.61¶
Wetting Angle	θ	degrees	21¦	20¦	66¥

* Cohen et al., 1993; ¶ 3M (1996); § Measured property (t=20-25°C); ¦ Measured property with Sudan IV (t=20-25°C); ¥ Measured property with red food colorant (t=20-25°C)

Fracture Plane Characterization. The fracture planes were characterized by three different equivalent apertures derived from hydraulic and tracer studies. For a set flow rate and measured hydraulic gradient, the hydraulic aperture, e_h, was determined using the cubic law given by

$$e_h = \left(\frac{12\mu QL}{\rho g W |\Delta H|} \right)^{\frac{1}{3}} \tag{2}$$

where μ is the kinematic viscosity, Q is the flow rate, L is the fracture length, W is the fracture width, ρ is the fluid density, g is the acceleration due to gravity, and ΔH is the head difference between the source and observation point.

The tracer tests involved the constant injection of a solution containing 84 mg/L solution of bromide, which is a conservative tracer. Both the influent and effluent flow cells were continuously mixed, and 1.5 mL samples were withdrawn from the effluent flow cell every 2 to 3 minutes for the duration of the experiment. Ion chromatography was used to determine the bromide concentration. The volume of the effluent flow cell was used in conjunction with the slope of the observed break-through curve to determine the concentration of bromide exiting the fracture plane into the flow cell.

The mass balance aperture, e_{mb}, which is based on the volumetric (plug flow) displacement, was determined from advective tracer data using the mean residence time (Tsang, 1992) as

$$e_{mb} = \frac{Qt_m}{LW} \tag{3}$$

where t_m is the mean residence time.

The frictional loss aperture, e_f, which is based on the difference in hydraulic head between the source and observation point, was derived by substituting the cubic law for Q in (3) to yield (Tsang, 1992)

$$e_f = L\sqrt{\frac{12\mu}{\rho g |\Delta H| t_m}} \tag{4}$$

RESULTS AND DISCUSSION

Visualization Experiments. Figure 2 shows the residual saturation of HFE 7100 in the artificial fracture plane. The residual mass is trapped by critically small aperture regions which the DNAPL was not able to penetrate even when flushed at a Ca as high as 2.5E-3. It is significant that the majority of the residual mass is concentrated in two large blobs. The effect of this mass configuration is that the interfacial area is small relative to the volume of mass entrapped; this configuration can be more closely compared to DNAPL pools in porous media than to residual saturations in porous media.

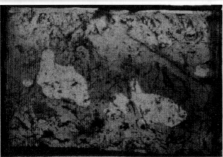

Figure 2. Residual saturation of HFE 7100 in the transparent fracture plane. The dark-colored phase represents the red-dyed water, and the light-colored phase represents the HFE-7100.

Aperture Field Characterization. The equivalent aperture data from the hydraulic and tracer studies is presented in Table 2. The relative magnitudes of the three equivalent apertures derived from the tracer and hydraulic studies are $e_{mb} \geq e_h \geq e_f$. These apparent differences are an artefact of the approach taken to the respective equivalent aperture calculations (Tsang, 1992), and are consistent with the findings from previous studies. By definition, e_{mb} represents the pore volume of the fracture and is therefore controlled by the largest aperture regions. Conversely, e_h and e_f are based on the pressure drop across the fracture plane, which is sensitive to the smallest aperture regions. It has been suggested that e_{mb} is the most worthy of the term 'equivalent aperture', as it represents an average along the flow paths of the tracer transport (Tsang, 1992). Since e_{mb} is sensitive to the largest aperture regions and e_f is sensitive to the smallest aperture regions, it follows that the ratio of e_{mb} to e_f can be used as a measure of the variability of the aperture field.

TABLE 2. Equivalent aperture data.

Technique	Fracture 1 (μm)	Fracture 2 (μm)	Synthetic Fracture (μm)	Error
Hydraulic	345	190	420	± 7%
Mass Balance	670	630	865	± 5%
Frictional Loss	275	112	260	± 10%
e_{mb}/e_f	2.44	5.62	3.33	

In most practical situations the residual NAPL distribution cannot be visualized, and therefore the interfacial area is not easily calculated. However, because the mass transfer rate is so sensitive to the interfacial area, it is important to be able to quantify the interfacial area in some way. Since the interfacial area is controlled by the entrapped NAPL configuration, which is in turn controlled by the variability of the aperture field, it follows that a measure of the variability of the aperture field may be a suitable surrogate measure for the interfacial area.

Dissolution Experiments. Figure 3 shows an effluent curve from a typical dissolution experiment (Trial 1-8). It can be seen that the effluent concentration peaked at fewer than 150 pore volumes, and slowly decreased as more pore volumes of water were flushed through the fracture plane. As the effluent concentration began to tail, the mass removal rate decreased until it was relatively constant. A summary of the results from the dissolution experiments (Table 3) indicates that at most 62% of the mass was removed after several thousand pore

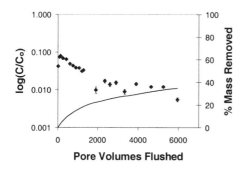

Figure 3. Example of an effluent concentration profile from Trial 1-8. The (♦) symbol represents log(C/C₀), and the (—) symbol is the % of mass removed.

volumes of water were flushed. These removal rates are significantly different from porous media removal rates reported in the literature for both residuals and pools. Column experiments performed with residual saturations typically achieve greater than 90% mass removal after several thousand pore volumes of water flushing (e.g. Powers et al., 1994). Model aquifer experiments studying the dissolution of pools in porous media typically see removal rates of less than 20% after several thousand pore volumes of water have passed through (e.g. Pearce et al., 1994).

Table 3. Results of the dissolution experiments.

Trial	DNAPL	Specific Discharge (m/day)	Pore Volumes Flushed	Mass Entrapped (g)	Mass Removed (%)
1-3	1,1,1 TCA	10.1	950	14.5	48
1-4	1,1,1 TCA	17.6	1915	15.7	60
1-7	1,1,1 TCA	24.0	4760	14.8	22
1-8	TCE	104.2	7255	11.48	19
1-9	1,1,1 TCA	48.0	6720	24.22	14
2-1	1,1,1 TCA	31.2	1800	32.1	40
2-2	1,1,1 TCA	70.6	2200	29	62

The difference in various mass removal rates observed for porous media and fractured media may be related to the differences in interfacial areas, since the dissolution rate is directly proportional to the interfacial area. The residual mass distribution within fracture planes appears to compare more closely to pools in porous media than to residuals in porous media.

CONCLUSIONS

Given the definitions of e_{mb} and e_f, the ratio of these two equivalent apertures provides a measure of aperture variability for a given fracture plane. This measure of variability may be a useful surrogate measure for the interfacial area, which is directly proportional to the mass transfer rate.

The percentage mass removals observed in these dissolution experiments were between typical percentage mass removals reported for pooled and residual sources in porous media. One potential explanation for this observed difference is the relative difference in interfacial areas between these mass distributions.

Ongoing work involves the application of non-linear regression techniques to the data collected throughout the dissolution experiments, and the development of a phenomenological model relating the mass transfer rate to flow velocity and interfacial area.

ACKNOWLEDGMENTS

This research was conducted with funding from the Natural Sciences and Engineering Research Council of Canada (NSERC), the University Consortium Solvents-In-Groundwater Research Program, and an NSERC post graduate scholarship awarded to the first author.

REFERENCES

3M. 1996. *3M HFE 7100 Product Information.*

Chatzis, I., N.R. Morrow and H.T. Lim. 1983. "Magnitude and detailed structure of residual oil saturation." *Soc. Pet. Eng. J.* 23(2):311-326.

Cohen, R.M., J.W. Mercer and J. Matthews. 1993. *DNAPL Site Evaluation.* C.K. Smoley, Boca Raton, FL.

Esposito, S.J. and N.R. Thomson. 1999. "Two-phase flow and transport in a single fracture-porous medium system." *J. Contam. Hydrol.,* 37:319-341.

Glass, R.J. and J.J. Nicholl. 1995. "Quantitative visualization of entrapped phase dissolution within a horizontal flowing fracture." *Geophys. Res. Letters*, 22(11): 1413-1416.

Murphy, J.R. and N.R. Thomson. 1993. "Two-phase flow in a variable aperture fracture." *Water Resour. Res.,* 24(12):2033-2048.

Parker, B.L., R.W. Gillham and J.A. Cherry. 1994. "Diffusive disappearance of immiscible phase organic liquids in fractured geologic media." *Ground Water*, 32(5): 805-820.

Pearce A.E., E.A. Voudrias, and M.P. Whelan. 1994. "Dissolution of TCE and TCA pools in saturated subsurface systems." J. of Env. Eng., 120(5):1191-1206.

Powers, S.E., L.M. Abriola, and W.J. Weber Jr. 1994. "An experimental investigation of nonaqueous phase liquid dissolution in saturated subsurface systems: Transient mass transfer rates." *Water Resour. Res.* 30(2):321-332.

Tsang, Y.W. 1992. "Usage of 'equivalent apertures' for rock fractures as derived from hydraulic and tracer tests." *Water Resour. Res.* 28(5): 1451-1455.

VanderKwaak, J.E. and E.A. Sudicky. 1996. "Dissolution of non-aqueous-phase liquids and aqueous-phase contaminant transport in discretely-fractured porous media." *J. Contam. Hydrol.,* 23:45-68.

DNAPL FLOW THROUGH FRACTURED POROUS MEDIA

David A. Reynolds and Bernard H. Kueper
(Queen's University, Kingston, Ontario, Canada)

ABSTRACT: A numerical sensitivity study has been carried out which focuses on the effects of matrix diffusion on the penetration of DNAPL through a single sub-horizontal fracture. The most significant physical parameters controlling the speed of migration were found to be the aperture and angle of inclination of the fracture, and the solubility of the DNAPL. It was also found that for large aperture fractures (greater than 30 μm) the combined effects of high non-wetting phase fluxes and large volumes within the fracture prevented matrix diffusion from significantly retarding the DNAPL front.

INTRODUCTION

The presence of DNAPL in the subsurface is recognized as a significant source of long-term groundwater contamination at many sites throughout North America and Europe (Cherry et al., 1996; Freeze and McWhorter, 1997). DNAPLs of environmental concern include PCB oils, chlorinated solvents, coal tar, and creosote.

A release of DNAPL to the subsurface will flow predominantly vertically under gravitational force until encountering a capillary barrier. Given sufficient quantities of release, flow will then become predominantly horizontal. In addition to migration horizontally, DNAPL will tend to pool above the capillary barrier, increasing the chances of further vertical migration by either overcoming the entry pressure of the barrier, or by entering large aperture imperfections such as fractures or macropores.

Kueper and McWhorter (1991) developed a conceptual model for DNAPL flow through fractured environments, pointing out that once entering a fracture network DNAPL will migrate preferentially through the largest aperture fractures and will in general prefer to migrate vertically downwards in response to gravity forces. Lateral flow will also occur, provided the entry pressure of intersecting horizontal fractures is overcome, if vertical migration pathways are incapable of accepting the incoming flux of DNAPL or are non-existent. DNAPL, due to its large relative density, will also enter dead-end vertical fractures, outside the connected wetting phase flow pathway.

Parker et al. (1994; 1996) pointed out that, due to the porous nature of the matrix surrounding fractures, the DNAPL trapped in fractures may dissolve into the aqueous phase and diffuse into the matrix. The life span of DNAPL in such situations can range from less than a day to several decades depending on the chemical composition of the DNAPL, the physical properties of the matrix, and the fracture aperture. Parker et al. point out that the disappearance times, which may be short relative to the ages of subsurface contamination at some sites, can result in DNAPL source zones containing little actual non-wetting phase after a sufficient period of time.

Ross and Lu (1999) analytically investigated the effects of mass transfer to the matrix on the advancement of DNAPL down a vertical fracture. Using a dimensionless "Nitao" number, Ross and Lu present a method of determining if the effects of matrix diffusion are significant on the rate of advance of a DNAPL front. The analytical formulation, though allowing for dynamic flow behaviour, did not incorporate the effects of relative permeability, or an explicit representation of capillary pressure gradients.

Numerical investigations of DNAPL flow in a single fracture (Slough et al., 1999, Esposito and Thomson, 1999) have examined the effects of porosity, aperture, and spacing on DNAPL migration, and on factors affecting mass removal from a fractured system. Slough et al. (1999) found that increasing the matrix porosity dramatically increased the required time for DNAPL to penetrate a 10m long fracture. This is consistent with the analytical solution of Parker et al. (1994) in which mass lost to the matrix is directly proportional to porosity. Slough et al.'s (1999) use of a constant non-wetting phase flux boundary condition at the top of the fracture, however, influences the relationship between matrix porosity and non-wetting phase front advancement.

The objective of this study is to advance the above work by considering fully the effects of gravity, capillary pressure, relative permeability, viscosity, and matrix diffusion on the rate of DNAPL migration through fractured porous media. This work is focused on a single fracture. Understanding at this scale is a prerequisite to understanding and analysis at the network and field scale.

THEORY AND SIMULATION DOMAIN

The model used in this work (QUMPFS) is a fully three-dimensional, multiphase compositional simulator, which allows for explicit simulation of matrix-fracture interactions, non-equilibrium phase partitioning, and hysteresis in the relative permeability and capillary pressure-saturations relationships.

The partial differential equations governing isothermal multiphase flow with multicomponent transport in porous media on which the model is based are:

$$\frac{\partial}{\partial t}\left(c_{i\beta}\phi\, S_{\beta}x_{i\beta}\right)+\nabla\cdot\left(c_{i\beta}x_{i\beta}v_{\beta}\right)-\nabla\cdot\left[\phi\, S_{\beta}\vec{D}_{i\beta}\nabla\left(c_{i\beta}x_{i\beta}\right)\right]-q_{i\beta}-I_{i\beta}=0 \quad (1)$$

$$\beta=1..n_{p}, \quad i=1..n_{c}$$

Where n_p is the number of phases, β is the phase of interest, n_c is the number of components, i is the component of interest, $c_{i\beta}$ is the molar density of phase β, S_β is the fraction of void space occupied by phase β, ϕ is the porosity of the medium, v_β is the flux of phase β, $x_{i\beta}$ is the mole fraction of component i in phase β, $D_{i\beta}$ is the dispersion tensor for component i in phase β, $q_{i\beta}$ is a source or sink of component i in phase β, and $I_{i\beta}$ represents the inter-phase mass transfer of component i to or from phase β. The governing equations are discretized through the finite volume method using a lumped mass time derivative.

The physical system used in this study is shown in Figure 1, and consists of a single fracture imbedded in a porous matrix. The fracture is 3m long, and is inclined from 90 to 70 degrees below horizontal in various simulations with unit depth into the third dimension. The spatial discretization perpendicular to the fracture grades from 100 microns adjacent to the fracture to over 10 cm at the edge of the domain. This fine discretization is required to handle the steep concentration gradient that exists at early time between the fracture and the matrix, and which greatly affects the behaviour in the system (Slough et al., 1999). The vertical discretiztion begins at 1 cm at the top (inlet) of the fracture, and grades to 10 cm at the bottom (outlet) of the fracture.

$P_c = 0.3$ m TCE

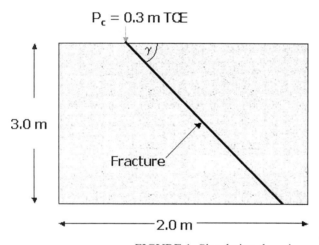

FIGURE 1. Simulation domain

Flow boundary conditions on the domain consist of no-flow along all vertical edges, and constant wetting phase pressure along the top and bottom of the domain, such that the pressure distribution is hydrostatic. Non-wetting phase was introduced into the top of the fracture under a constant capillary pressure of 4295 Pa, equal to 0.3 m of pooled TCE situated at the water table, which is located at the top of the domain. This value was chosen to allow entry into all fracture apertures simulated in this study. All simulations terminated when non-wetting phase appeared in the lowest cell in the fracture. Transport boundary conditions were set such that aqueous phase contaminant was unable to leave the domain.

Table 1 details the parameters used in the base-case simulation, and highlights the specific parameters that were varied during this sensitivity study. The choice of parameters to vary in the study (aperture (e), solubility (S), porosity (ϕ), distribution coefficient (K_d), and fracture inclination (γ)) was based on the work of Kueper and McWhorter (1991) and Parker et al. (1994), who identified these as the key parameters controlling migration. Table 2 presents the ranges of values used for each of the key parameters. In general, the geological parameters were kept within what could be considered appropriate for a clay or silty-clay deposit. The DNAPL solubility was increased from that of pure TCE (all other NAPL properties are representative of TCE) to that of 1,1-Dichloroethane.

RESULTS AND DISCUSSION

To provide a framework for the discussion of the results, the base-case was initially run using a variety of fracture apertures both with and without allowing for DNAPL dissolution and matrix diffusion. The non-dissolution/diffusion case represents the fastest possible migration through the fracture. Table 3 summarizes the breakthrough times. For apertures ranging from 10 to 75 microns, the time to traverse the entire fracture ranges from 11.3 days to 24.5 minutes for the non-dissolution case, and from 13.7 days to 25 minutes for the dissolution case. It is clear that matrix diffusion slows the rate of migration through the fracture, but not by a significant amount.

TABLE 1. Physiochemical parameters utilized in base-case

Parameter	Value	Parameter	Value
Pore Size Distribution Index	2.0	Matrix Permeability	$1.0 \times 10^{-17} \text{ m}^2$
Residual Wetting Phase Saturation	0.1	Longitudinal Dispersivity	0.01 m
Wetting Phase Viscosity	$1.0 \times 10^{-3} \text{ Pa s}$	Transverse Dispersivity	0.001 m
Non-wetting Phase Viscosity	$5.7 \times 10^{-4} \text{ Pa s}$	*Aperture*	10 μm
Diffusion Coefficient	$2.0 \times 10^{-9} \text{ m}^2/\text{s}$	*Solubility*	1325 mg/L
Wetting Phase Density	1000 kg/m^3	*Matrix Porosity*	0.1
Non-wetting Phase Density	1460 kg/m^3	*Distribution Coefficient*	$1.26 \times 10^{-4} \text{ L/kg}$
Interfacial Tension	$2.0 \times 10^{-2} \text{ N/m}$	*Fracture Inclination*	90 °

Direct comparison with the work of Ross and Lu (1999) is difficult due to the assumptions inherent in their analytical solution procedure. This is most evident when comparing the predicted rates of advance of the non-wetting phase through the fracture. Ross and Lu (1999) use a formulation that requires two simplifications of the physics involved with two phase flow; neglecting over-pressurization of the system by a capillary pressure at the fracture entrance greater than the displacement pressure, and the concept of relative permeability.

TABLE 2. Values of parameters varied between simulations

Parameter	Values
Aperture	10, 20, 30, 50, 75 μm
Solubility	1325, 2000, 5000 mg/L
Matrix Porosity	0.1, 0.3, 0.5
Distribution Coefficient	1.26×10^{-4}, 6.3×10^{-4}, 1.26×10^{-3} L/kg
Fracture Inclination	90°, 80°, 70°

These two effects are offsetting, however, and their relative importance cannot be assessed without the use of a numerical model. In the 10 μm fracture simulation, the difference resulting from how capillary pressure is treated between the analytical and numerical formulations is small, as the capillary pressure at the entrance of the fracture is only 7% greater than the displacement pressure of the fracture. The relative permeability of the fracture to DNAPL, however, is low at the saturation corresponding to a capillary pressure of 4295 Pa (3.1×10^{-4} Brooks and Corey, 1964). This results in the increased breakthrough times presented in Table 3 as compared to that predicted by Ross and Lu (1999) for parameters identical to the base-case (note this does not include the excess capillary pressure applied at the entry of the fracture). Table 3 clearly shows that fracture aperture is much more influential than matrix diffusion in dictating the rate of migration through the fracture.

TABLE 3. Breakthrough times for base-case simulations

Aperture (μm)	Breakthrough Time (Ross and Lu, 1999)	Breakthrough Time (No Dissolution)	Breakthrough Time (With Dissolution)
10	12.6 hours	11.3 days	13.7 days
20	3.2 hours	6.7 hours	8.4 hours
30	1.4 hours	2.8 hours	3.5 hours
50	30.3 minutes	56.6 minutes	63.6 minutes
75	13.5 minutes	24.5 minutes	25.6 minutes

Figure 2 shows the complete results for the sensitivity study, plotted as the ratio of breakthrough time for the simulation with the altered parameter to the base-case simulation (with dissolution). The effects of all of the parameters are small to moderate when compared to the effect of fracture aperture, with less sensitivity for larger aperture fractures. The two most important parameters controlling the advance of the non-wetting phase found with this study, independent of fracture aperture, are the solubility of the DNAPL and the angle of the fracture. The effect of increasing the solubility of the DNAPL directly increases the mass lost from the fracture due to matrix diffusion.

Parker et al. (1994) express the mass diffused into the matrix at time t, per unit area of fracture, as:

$$M_c = \phi \, S_w \, \frac{4}{\sqrt{\pi}} \sqrt{RD^* \tau \, t} \tag{2}$$

The results of this study show that the linear relationship between the mass lost due to diffusion into the matrix and porosity in equation 2 is evidenced by a near-linear relationship between breakthrough time and this parameter. This result was also reported by Slough et al. (1999), though for a dissimilar boundary condition. In addition, it is of interest to note that a 500% increase in the matrix porosity does not retard the DNAPL front as significantly as a 377% increase in the solubility. This is due to the significant increase in sorbed mass that occurs in conjunction with the increase in solubility. The importance of sorption in the system is further evidenced with f_{oc} being the third most important (independent of aperture) factor controlling DNAPL flow.

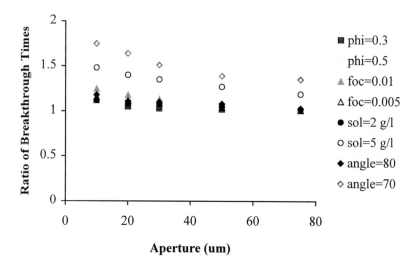

FIGURE 2. Summary of results

The factor exhibiting the most pronounced effect on the migration of the DNAPL (independent of fracture aperture) is the angle of inclination of the fracture. A decrease of 20° causes an approximate doubling in the breakthrough time (for the 10 μm fracture). Reducing the incline of the fracture is the only change made to the system that significantly alters the quantity of non-wetting phase entering the domain. The gravitational force on the flowing DNAPL is proportional to the sine of the angle of inclination, thus even steeply dipping (but not vertical) fractures can experience a significant reduction in the volume of non-wetting phase flow.

A striking feature of the behaviour summarized in Figure 2 is the smaller difference that varying the parameters makes on fractures larger than 30 μm. This phenomenon is due to a combination of the increased capillary driving force and the larger volumes of DNAPL per square metre of fracture. The reduced time required for the DNAPL to traverse the larger fractures under the greater capillary gradient and higher DNAPL relative permeability reduces the time available for diffusive losses to the matrix. In addition, the diffusive losses become a much smaller percentage of the mass available with an increase in fracture aperture. In recognizing that these two parameters are additive, it is doubtful that breakthrough times in larger aperture fractures will be significantly reduced in any realistic setting with any low solubility organic compound.

The balancing of the mass lost through matrix diffusion and the mass entering the fracture is the overall controlling factor on the rate of DNAPL penetration through the system. In a macroscopic sense, if the mass flux to the matrix exceeds the mass flux into the fracture, then vertical migration may stop (though this was never seen in this work). In certain cases, where the diffusive flux into the matrix exceeds the non-wetting phase advective flux into the top of the fracture, downward migration may still occur due to gravity forces, but with the system experiencing capillary hysteresis (reduction in DNAPL saturations while still maintaining downward flow).

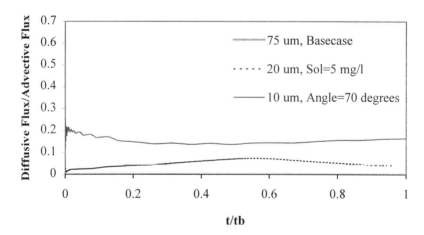

FIGURE 3 – Ratio of mass fluxes for various options

Figure 3 presents the ratio of diffusive flux to that of advective flux for several simulations. When the ratio plotted on the ordinate axis exceeds 1.0, there is a net loss of mass from the fracture. This did not occur for any of the simulations presented here, as in all cases the majority of the mass remained within the fracture in pure phase form. The simulation with the greatest loss ratio (10μm aperture, 70° inclination) also exhibited the greatest increase in breakthrough time. Two other cases are presented in Figure 3, an intermediate case where the diffusive flux again never exceeds the advective flux, but slightly retards the non-wetting front (20 μm aperture, solubility of 5 g/l), and a case where the loss of mass does not affect the system to any significant degree (75 μm base-case).

REFERENCES

Brooks, R.H. and A.T. Corey. 1964. "hydraulic Properties of Porous Media." *Hydrol. Paper 3*, Colorado St. Univ., Fort Collins CO.

Cherry, J.A. S. Feenstra, and D.M. McKay. 1996. "Concepts for the Remediation of Sites Contaminated with Dense Nonaqueous Phase Liquids (DNAPLs)." In J.F. Pankow and J.A. Cherry (Eds.), *Dense Chlorinated Solvents and Other DNAPLs in Groundwater: History, Behaviour, and Remediation, pp.* 475-506, Waterloo Press, Portland, Oregon.

Esposito, S.J., and N.R. Thomson. 1999. "Two-phase Flow and Transport in a Single Fracture-porous Media System." *J. Cont. Hyd.* 37: 319-341.

Freeze, R.A., and D.B. McWhorter. 1997. "A Framework for Assessing Risk Reduction Due to DNAPL Mass Removal from Low Permeability Soils." *Ground Water.* 35(1): 111-123.

Kueper, B.H., and D.B. McWhorter. 1991. "The Behavior of Dense, Nonaqueous Phase Liquids in Fractured Clay and Rock." *Ground Water.* 29(5): 716-728.

Parker, B.L., R.W. Gillham, and J.A. Cherry. 1994. "Diffusive Disappearance of Immiscible-Phase Organic Liquids in Fractured Geologic Media." *Ground Water.* 32(5):805-820.

Parker, B.L., D.B. McWhorter, and J.A. Cherry. 1996. "Diffusive Loss of Non-Aqueous Phase Organic Solvents from Idealized Fracture Networks in Geologic Media." *Ground Water.* 35(6): 1077-1088.

Ross, B., and N. Lu. 1999. "Dynamics of DNAPL Penetration into Fractured Porous Media." *Ground Water.* 37(1): 140-147.

Slough, K.J., E.A. Sudicky, and P.A. Forsyth. 1999. "Numerical Simulation of Multiphase Flow in Discretely-fractured Geologic Media." *J. Cont. Hyd.* 40(2): 107-136.

REMEDIATION OF SMITHVILLE, CANADA, PCB SPILL SITE

J. E. (Ted) O'Neill (Smithville Phase IV, Smithville, Ontario Canada)
John Mayes (Ministry of the Environment, Guelph, Ontario Canada)

ABSTRACT: The Smithville Phase IV Bedrock Remediation Board was established to manage and remediate a site in Smithville, Ontario, Canada, where 30,000 L of DNAPL (consisting of PCB, TCB and TCE) leaked into a fractured carbonate rock aquifer. The Board established a team of world-class experts and formed partnerships with research groups in Canada and the United States to develop new technical solutions for the contamination. The experts successfully characterized the site using state-of-the-art methods. The engineering studies are evaluating four groups of alternatives: a permeation grouted barrier in the fractured bedrock to contain the aqueous plume, pump and treat, monitored natural attenuation and thermal wells to remove the source zone. Combinations of these technologies were also under consideration.

INTRODUCTION

Polychlorinated Biphenyls (PCBs) were used extensively in electrical equipment for decades before they were found to pose a serious environmental risk and banned from general use in the 1970's. Their persistent, robust, almost indestructible nature made PCBs a good choice as an electrical insulator in hydro transformers, capacitors and other types of electrical equipment. Unfortunately, these same characteristics also make the safe and efficient management of PCB waste a challenging problem.

In the past decade, significant progress has been made in several areas dealing with the recovery and destruction of PCBs spilled in the environment. Today there are safe and efficient ways to destroy PCB wastes held in storage or recovered from spill sites. The challenge is no longer how to destroy PCBs but how to effectively recover spilled wastes from the environment. Experience has shown that mass removal rates of PCBs and other contaminants known as Dense Non-Aqueous Phase Liquids (DNAPLs) from the environment may have to exceed 99.9% or more, otherwise groundwater contamination may still exceed drinking water objectives.

Just ten years ago there were no proven technologies for remediating any PCB spill sites. This is no longer the case. Recent research and technology development efforts in DNAPL have focused on porous media research. Today there are proven and emerging technologies capable of dealing with PCB in porous media, such as: in-situ thermal desorption, steam flushing, solvent extraction and permeable reactive barriers.

Fractured media flow and contaminant transport on the other hand are recognized as much more complex issues and consequently have received comparatively little attention until recently. At Smithville, Ontario, a PCB spill site (known as the former CWML site) has been the subject of extensive, state-of-the-science site characterization work and a 10-step decision-making process to identify a preferred remediation alternative(s).

This paper describes the progress made in finding and evaluating suitable technologies for remediating PCBs and other contaminants in the bedrock at the Smithville site.

BACKGROUND

In 1994, a Board of Directors was established to manage and implement a program to contain and remediate the former CWML site in Smithville. PCB contamination in the environment was discovered in late 1985. It has been estimated that approximately 30,000 L of DNAPL (consisting of PCB, TCB and TCE) leaked from the waste transfer station and penetrated the fractured carbonate rock beneath the site. A safe water supply was provided by the Ontario government to replace Smithville's threatened groundwater supply and a pump and treat system was installed to prevent off-site movement of dissolved phase contaminants. The pump and treat system has worked effectively to contain the dissolved phase plume since it was installed in 1989; however, the rate of removal of the DNAPL source is extremely low. The objective of the Board is to identify cost-effective, long-term alternatives for remediation of the remaining DNAPL in the bedrock and groundwater beneath the site.

Recognizing the significant challenge posed by DNAPL in fractured media the Smithville Board established a team of world-class experts and formed partnerships with research groups in Canada and the United States to develop new technical solutions. Between 1994 and 1999, the Board's team of experts successfully characterized the site using state-of-the-art methods, conducted laboratory studies to characterize the bedrock and DNAPL properties, completed modelling work to investigate the possible movement of DNAPL in bedrock and undertook engineering studies to evaluate and select remediation technologies for possible application at the Smithville site.

IDENTIFICATION OF ALTERNATIVES

By March 1998, the Board and its technical team considered over 40 potential remedial alternatives for possible application at Smithville. Using criteria established for the decision-making process (Smithville, 1997) many alternatives were screened from further consideration because they were inappropriate for the site conditions and/or contamination.   Nine alternatives survived the screening step, as follows:

Mass-Removal Alternatives
-   excavation and ex-situ treatment
-   thermal wells

Migration Control
-   ground freezing

- hydromill excavating
- secant piling
- permeation grouting
- extraction wells
- integrated permeation grouting and extraction wells.

Natural-Attenuation Alternative
- monitored natural attenuation (MNA)

In 1998, Acres and Associated Limited was retained to provide the basis for a further narrowing of the alternatives. The engineering work was undertaken in two stages. The first stage was to prepare a *Common Design Basis Report* to document, from an engineering perspective, the important features of the site and contaminants in regard to remediation alternatives. The second stage was to prepare pre-feasibility level design descriptions to serve as a basis for a detailed comparative evaluation of the alternatives.

A fundamental requirement in the Board's decision-making process was to establish a realistic site conceptual model to serve as a basis for identification, evaluation and recommendation for a preferred alternative(s). Site characterization work at Smithville was undertaken by a team of experts led by Dr. Kent Novakowski, Brock University and was completed in 1999. The work of Dr. Novakowski and his team established a unique, research-quality database for the site that has gained recognition in the international scientific community. Smithville is currently regarded as the best characterized DNAPL-contaminated fractured rock site anywhere. On the basis of this leading-edge site characterization work the team developed a conceptual model for groundwater flow in the carbonate bedrock at the site (Novakowski *et al.*, 1999).

The first task facing the engineering team was to consider the site conceptual model developed by the scientists and to ensure that it was sufficient for engineering purposes. Data gaps were identified and a set of common design features or characteristics pertaining to general site conditions and assumptions regarding the site were documented (Smithville, 1999). The primary focus of the Common Design Basis was on geology, hydrogeology, free-phase DNAPL contamination, aqueous-phase contamination and potential receptors.

The distribution of DNAPL in the bedrock is extremely varied and complex, and concentrations can vary substantially among different fracture horizons and even within a fracture. Due to the heterogeneity of the bedrock and the technical difficulty of determining the location of the DNAPL, estimates can be subject to considerable uncertainty. Therefore, the exact limits of the horizontal and vertical distribution of the contaminant mass remain unknown. However, scoping calculations indicate that the bulk of the original contaminant mass is present in the subsurface, within the immiscible and sorbed phases (MacFarlane, 1996).

Modelling using site characterization data has shown that the DNAPL moved through the overburden underlying the former lagoon to the bedrock surface. Upon reaching the bedrock, the DNAPL then moved along the surface until a sufficiently large fracture aperture was encountered. Vertical fractures on the bedrock surface would provide a path for the material to move down the joints until a dead-end or the fracture terminated at a horizontal bedding plane. The DNAPL would then move laterally along the path of least resistance until it intersected with another vertical fracture. Tortuous pathways through vertical fractures interconnected to horizontal fractures may have provided a "step-wise" pathway for DNAPL to move deeper into the bedrock.

DNAPL occurrence was identified in bedrock cores and in groundwater samples during various investigations at the site. Recently, a more detailed assessment of the extent of the DNAPL source was completed (Golder, 1999). Geochemical sampling and testing suggests that the majority of the DNAPL is within an area approximately 200 m wide in an east-west direction, and 250 m long in a north-south direction, along the direction of groundwater flow.

The depth of DNAPL penetration into the bedrock is controlled by both the site geology and the contaminant characteristics. The interconnectivity of the fracture network, especially in the vertical plane, is a key element in the eventual distribution of the DNAPL. It is impractical to measure or even estimate all of these factors, however, numerical simulations can be used to evaluate some of these factors and the implications at the site.

Monitoring in the source area is limited to the Upper Eramosa so that the vertical depth of DNAPL contamination is not known. However, DNAPL has penetrated at least to a depth of the monitoring points, which is approximately 11 m below the ground surface. Results from recent investigations have indicated that there has been no significant migration of DNAPL below the Lower Vinemount (Golder, 1999; Novakowski *et al.*, 1999).

The aqueous-phase plume once extended up to 800 m to the downgradient of the former lagoon. However, the plume has reduced in size as a result of operation of the on-site pump and treat system that was installed in April 1989. This plume consists of largely TCE and TCB contamination. This contamination is primarily produced through contact with DNAPL compounds.

Groundwater in the Upper and Lower Eramosa contain the highest aqueous-phase concentrations. This zone of aqueous-phase contamination constitutes the greatest volume of contamination, however, the actual mass of contaminants in the aqueous phase is small. For example, TCE in the DNAPL accounts for approximately 2% of the contaminant mass, but it is the most mobile contaminant detected in the aqueous-phase plume, downgradient of the site due to TCE's relatively high solubility in water, compared to the other DNAPL constitutes.

FEASIBILITY LEVEL DESIGN DESCRIPTIONS (STAGE 2)

The purpose of the Design Descriptions was to provide technical and financial information to compare the various alternatives. Information from other steps in the process provided environmental and community information used in the selection of a preferred alternative(s). In 1999, the Board used the Stage 2 work as a basis for a further narrowing of the alternatives. This involved consideration of technical and financial information from the *Common Design Basis Report*, unit flow diagrams illustrating the implementation steps for each alternative and discussions with the engineering team regarding schedule, technology experience, operational life of the system, monitoring requirements, costs including capital, operation and maintenance and effects on adjacent properties.

For the purpose of discussion herein, four representative remediation alternatives are presented. These alternatives, representative of the primary groups, are: a physical barrier constructed in the fractured bedrock by permeation grouting to contain the aqueous phase plume, extraction wells and a treatment system to contain the aqueous phase plume, monitored natural attenuation and thermal wells to remove the source zone. Combinations of these technologies were also under consideration.

Monitored Natural Attenuation. Monitored natural attenuation (MNA) utilizes the natural subsurface processes of biodegradation, sorption, dilution, dispersion, volatilization, and chemical reactions to reduce contaminant concentrations to acceptable levels. Monitored natural attenuation is an alternative that combines thorough site characterization, predictive modelling, risk assessment, and long-term monitoring to determine if the natural processes can assimilate the contamination present to an extent that is sufficiently protective of human health and the environment.

In the United States, MNA has been selected at Superfund sites where, for example, PCBs are strongly sorbed to deep overburden and not migrating, where removal of DNAPL is technically impractical, and where active remediation measures would not sufficiently speed up the remediation time frame (EPA, 1993).

Mass Removal Alternative - Thermal Wells. This alternative would comprise in-situ thermal desorption (ISTD), using thermal wells to reach contamination in the bedrock beneath the site. ISTD is a mass removal technology that destroys most of the contamination below ground. The contaminants are either pyrolyzed *in-situ* or vaporized and subsequently captured using a vapor extraction system. This technology has been used successfully on various sites and with many different contaminants, including PCBs (Vinegar *et al.*, 1997).

Contaminants are removed by vaporization, in-situ thermal decomposition, oxidation, combustion and steam distillation. Thermal oxidation of the mobilized contaminants occurs during the residence time in materials adjacent to the wells as the vapors are pulled through the bedrock and out of the well casing by the vacuum applied to the well head. To successfully remove high boiling point contaminants, such as PCBs, the thermal wells must heat the subsurface contaminated zones to temperatures in excess of 300?C. Therefore, all the groundwater in the geologic unit being treated must be removed, either by boiling or a combination of pumping and boiling. Containment of the bedrock volume will be needed if the rate of groundwater influx into the treated regions is likely to exceed the vaporization capacity of the ISTD process. Because of the potential for high groundwater flows into the treated volume, it is likely that some barrier technology along with groundwater extraction would be necessary to effectively implement the technology.

Migration Control Alternative – Permeation Grouting. This application would involve surrounding the DNAPL source zone with a low permeability grouted barrier. The barrier would extend to a low hydraulic conductivity formation which would form the base of the containment area.

Permeation grouting involves a process where either a regular or an ultrafine cement based suspension grout is injected into the formation, to completely fill fractures, crevices and joints in the bedrock. For very fine fissures, which may be inaccessible, or only partly accessible, to ultrafine cement-based grouts, or for intergranular pores, permanent solution grouts are used to reduce the residual permeability.

Utilization of state-of-the-art grouting technology, with "real-time" monitoring and adjustment of the formulation of additivated, balanced regular and ultrafine cement-based suspension grouts, produces a less expensive, lower permeability and more durable grout curtain than past "conventional" grouting techniques (Dreese & Wilson, 1998). After completion of an ultrafine suspension grouting program, the residual permeability of the grouted rock can be further reduced by two orders of magnitude by secondary injection of water reactive, hydrophobic polyurethane grout (Anderson, 1998).

Techniques which have been established in the oil field to enhance formation penetrability could be applied to the carbonate bedrock at the Smithville site. Acid stimulation is one of the more prominent ones; which has been successfully applied in constructing cut-off curtains with extremely low residual permeability (less than 0.01 Lugeon; approximately equivalent to a hydraulic conductivity of 1×10^{-7} cm/s).

Since this alternative only provides a barrier, it is presumed that some groundwater extraction and/or capping technology would also be necessary to reduce or minimize the hydraulic head within the containment area.

Extraction Wells. This remedial alternative involves extraction of groundwater via wells to induce inward hydraulic gradients and form a groundwater capture zone. Extracted water must be treated prior to discharge, giving the technology its popular name, 'pump and treat'. Although contaminants are removed during extraction, this technology is viewed as a migration-control alternative rather than a mass-removal alternative as the system captures aqueous-phase contaminants without removing significant contaminant mass.

As the selection process moves forward the modelling, and risk assessment of the alternatives is proceeding concurrently. This information will contribute to the process and the selection of a preferred alternative or a combination of alternatives.

SUMMARY
The Board is in the final stages of its decision-making process and expects to make its recommendation to the Ministry of the Environment later this year.

ACKNOWLEDGEMENTS
The work described herein was funded entirely by the Ontario Ministry of the Environment. The contributors of the following are gratefully acknowledged: Messrs. Warren Hoyle and Carl Bodimeade, Acres & Associated Environmental Limited, and Mr. Larry Zamojski, Acres International Corporation.

REFERENCES

Anderson. 1998. "Chemical Grouting – An Experimental Study on Polyurethane Foam Grouts." Ph.D. Thesis, Tekniska Hogshola, Goteborg, Sweden.

Dresse, T.L., and Wilson D.B. 1998. "Grouting Technologies for Dam Foundations." In proceedings of ASCE Grouting Conference (October), Boston, MA.

Golder Associates Limited. 1999. *Assessment of Extent of PCB DNAPL Plume at CWML Site, Smithville, Ontario.* Prepared for the Smithville Phase IV Bedrock Remediation Program. (August).

MacFarlane, S., 1996. *Fate of Contaminants Analysis Study, Former CWML Site, Smithville, Ontario.* Prepared for Smithville Phase IV Bedrock Remediation Program, September, Site Archive No. R350, p.47.

Novakowski, K., Lapcevic, P., Bickerton, G., Voralek, J., Zanini, L. and C. Talbot. 1999. *The Development of a Conceptual Model for Contaminant Transport in the Dolostone Underlying Smithville, Ontario.* Prepared for the Smithville Phase IV Bedrock Remediation Program. Final Draft (November).

Smithville Phase IV Bedrock Remediation Program. 1997. *Process Overview Document.*

Smithville Phase IV Bedrock Remediation Program. 1999. *Engineering Feasibility Study, Stage 1, Common Design Basis Report.* Final Draft (September 13).

U.S. Department of Defense. 1994. *Remediation Technologies Screening Matrix and Reference Guide - Second Edition.* Prepared by the Department of Defence Environmental Technology Transfer Committee and Federal Remediation Technologies Roundtable. EPA/542/B-94/013.

Vinegar, H. *et al.* 1997. "In-situ Thermal Desorption (ISTD) of PCBs." In proceedings of Hazwaste World/Superfund 17 (October 1996).

A COMPARISON OF OPEN EXCAVATION AND BOREHOLE TECHNIQUES FOR ASSESSING DNAPL MIGRATION

Eric J. Raes, P.E. (Environmental Liability Management, Inc., Princeton, NJ)
Mark King, Ph.D. (Groundwater Insight, Nova Scotia, Canada)
James Cook, P.E. (Beazer East, Inc., Pittsburgh, PA)
Kenneth J. Luperi, P.G. and Hank Martin, P.E. (ELM, Inc., Princeton, NJ)

ABSTRACT: Test pit and borehole investigations were performed to assess Dense Non-Aqueous Phase Liquid (DNAPL) distribution and transport mechanisms at a former coal tar coatings facility in New Jersey. A consortium of organic constituents was historically released into glacial till overburden at the site. These constituents consisted primarily of the Chlorinated Aliphatic Hydrocarbons (CAHs), tetrachloroethene (PCE), and trichloroethene (TCE). They also included monoaromatics and polycyclic aromatic hydrocarbons (PAHs). PCE and TCE degradation products, including cis-1,2-dichloroethene (cis 1,2-DCE) and vinyl chloride (VC), are also present. All these constituents are present in both DNAPL and dissolved phases, and they occur in both the glacial till and in the underlying bedrock.

A comparison of results from two different subsurface investigation methods (i.e., boreholes and test pits) illustrates the differences that these methods may produce when interpreting DNAPL movement. In general, the direct observation of the glacial till that was made possible by the test pits provided substantially greater insight into DNAPL migration processes and distribution.

INTRODUCTION

The migration and distribution of DNAPL in the subsurface are strongly influenced by small- and large-scale variations in geologic structure. Consequently, the spatial distribution of DNAPL often is highly irregular (e.g., Johnson and Kueper [1996], Yu [1995]). In stratified unconsolidated deposits, much of the DNAPL mass may be present in thin horizontal accumulations on less permeable strata and as vertical stringers through more permeable materials.

DNAPL delineation in the subsurface is typically conducted with borehole investigations, since the depth of many DNAPL occurrences will preclude test pit investigation. Borehole investigations are commonly laborious and costly, and they are known to have significant limitations for interpreting DNAPL migration and distribution. This is due to the inherent difficulty in visualizing what may be a convoluted three-dimensional entity with information collected along vertical probes that are essentially one-dimensional.

With these limitations in mind, two borehole investigations were conducted at a site in New Jersey, to collect the necessary information to develop a site conceptual model for DNAPL migration and distribution. The initial investigation was used to identify the relative location of gross contamination, and was followed by a more focused program where centimeter-by-centimeter inspection of the soil cores was performed, with the visual aid of UV fluorescence. Soil and ground water samples were collected for laboratory analysis based on the unaided and aided visual observations.

The borehole investigations at the subject site were followed up with a test pit investigation that provided the opportunity to compare earlier borehole interpretations with those from test pits advanced in the same zone. In comparison with boreholes, test pits provide an enhanced capability to observe two- and three-dimensional characteristics of geologic media and DNAPL distributions, although the penetration depth of the latter was limited.

SITE BACKGROUND

History. The subject facility produced paints, coal-tar enamel, and related coal-tar products from the 1920s until operations ceased in 1994. Coal tar epoxy and other tar coating processes were added to the product line in the early 1970s.

The raw materials consisted of monoaromatic hydrocarbons (primarily toluene and xylenes) and coal tars containing PAHs. However, TCE and PCE were used both as formulation products and equipment cleaning agents. PCE was reportedly discharged into a 2- by 2-ft (.6 by .6 m) dry well located within the facility tank farm, as both frequent small losses during equipment cleaning and infrequent larger releases of pure product during materials formulations. The cleaning materials losses are suspected to have contained dissolved aromatic hydrocarbons and PAHs, which were also present in product formulations. However, it is possible that separate releases of these materials to the subsurface were later dissolved by releases of pure phase PCE.

Geology. The site is underlain by 32 to 37 ft (10 to 11 m) of glacial till (unstratified) and outwash (stratified) deposits. The till includes both ground moraine basal till and terminal moraine deposits. Both till types are well-graded, consist primarily of clay and silts, and are dense, with low permeability. Glacial outwash deposits are intercalated with the till, in an irregular sequence. The outwash deposits consist of well-sorted sand; layers of this unit range up to 4 ft (1.2 m) in thickness, but are typically less than 1 ft (0.3 m) thick. Shale and mudstone bedrock (Passaic Formation) occurs at 32 to 37 ft (10 to 11 m) and is typically highly weathered for the top 15 ft (5 m). In summary, the overburden at the site consists of an intercalated sequence of low permeability till and irregularly distributed higher permeability sandy zones.

BOREHOLE INVESTIGATIONS

A preliminary characterization of the horizontal extent of the CAHs and aromatic hydrocarbons in overburden was conducted in 1996. This investigation used a traditional outside-in approach and consisted of the installation of 15 borings and collection of 63 soil and 24 ground water samples. Based upon the results of visual observations, field screening results and analytical data, the initial extent of residual DNAPL was estimated and a shallow DNAPL source zone was identified in the vicinity of the dry well in the tank farm. Vertical delineation was terminated prior to the top of the underlying bedrock, to preclude the possibility of mobilizing DNAPL downward into bedrock. The investigation was terminated generally at 12 to 14 ft (4 to 5 m) below ground surface, whereas bedrock is encountered at 32 to 37 ft (10 to 11 m).

A second borehole investigation was performed to further characterize DNAPL vertical extent and migration pathways within the shallow source zone. Soil samples obtained from the boreholes were evaluated for DNAPL by visual observation, UV florescence and laboratory testing for CAHs. Additionally, a selected sample (i.e., SB-2) was evaluated for these parameters and PAHs. Through this investigation, it was concluded that the DNAPL had penetrated the entire thickness of the soil overburden. This conclusion was further supported by the observation of DNAPL in bedrock cores (using UV florescence) collected during a subsequent (third) borehole investigation. Table 1 provides a general indication of relative concentrations of various constituents in the DNAPL encountered at the site. These data are from the two overburden samples that contained representative concentrations of PCE and other constituents of concern.

TABLE 1: Concentrations (mg/kg soil) of selected organic constituents in overburden samples.

Constituents	Sample SB-3	Sample SB-2
PCE	53	37
TCE	0.470	--
Cis-1,2 DCE	0.810	--
VC	--	--
Xylene	1.6	--
Naphthalene	0.550	0.34
Other Aromatics	--	13

Analyte not detected.

TEST PIT INVESTIGATION

Subsequent to the borehole investigations, a test pit program was conducted to provide a more direct observation of shallow subsurface conditions. It was initially planned that DNAPL would be identified in the sidewalls of the test pits with the use of UV florescence and Sudan IV dye shaker tests. However, these enhanced observation methods were not required because the DNAPL was a dark gray color and readily identifiable by direct visual observation.

Test pits were excavated to depths of approximately 20 ft (6 m). Examination of test pit sidewalls and floors provided the opportunity for direct physical examination of stratigraphy and geological structure. Fracturing within large clasts (e.g., cobbles and boulders), lateral continuity of fine-grained layers, and changes in permeability could also be easily identified in the test pits. Such features could not be discerned with a high degree of confidence from previous soil borings.

Examination of the test pits also allowed the evaluation of DNAPL distribution and the relationship between the geological structure and the DNAPL migration.

COMPARISON OF RESULTS

The distribution of DNAPL in the subsurface is strongly influenced by the timing of the release, the volume of the release, and the heterogeneity of the porous medium into which the release occurs. Examination of the test pits identified substantial variability in the distribution of DNAPL throughout the source area. Characterization based only on borehole information would have provided a substantially different conceptual model of DNAPL distribution and migration processes. The following specific differences were identified in DNAPL conceptualization on the basis of test pit and borehole information.

- The test pits provided a clearer understanding of the transition from bulk DNAPL flow to discrete flow in fractures. Bulk DNAPL flow was observed within the more permeable sediments, but transitioned to discrete flow when it encountered low permeable sediments. The transition points were clearly evident within the test pits, but were typically not evident within the soil samples from the borings (i.e., either bulk flow or discrete flow was observed within the boring samples).

- Test pits provided a much higher degree of confidence in concluding where free phase DNAPL was present in the shallow overburden. The test pit observations provided a straightforward assessment for the presence of free phase DNAPL. The soil boring observations underestimated the areal extent of the free phase DNAPL in the shallow overburden. However, since only a small portion of the overburden was observable within the split spoon samples and the overburden was highly heterogeneous, it was recognized that the occurrence of free–phase DNAPL could be underestimated by relying only on soil boring samples. The test pit observations confirmed this prediction.

- The test pits allowed identification of previously unknown geological factors influencing DNAPL migration. The degree of DNAPL migration through relatively macroscopic fractures in the till had been expected and was confirmed by the soil boring samples. However, the test pit observations showed that DNAPL migration occurred, at times to a significant extent, along microfractures in the till. Observations of the soil boring samples had alluded to this potential, but the occurrences of DNAPL in boring sample microfractures was inconsistent and capricious, thereby precluding an accurate assessment of DNAPL extent. Test pit observations allowed fracture continuity, including microfractures, to be observed on a scale where pragmatic remediation decisions could be made with confidence.

• Observations from the test pits indicated that DNAPL was present outside the "source zone", as previously characterized. The presence of the microfractures and the degree of their continuity likely served as a major pathway responsible for this occurrence. It is thought that these microfractures resulted primarily from matrix desiccation. The boring sampling likely failed to detect the majority of these occurrences because of their predominant vertical/subvertical orientation. Observations of the test pit floors and walls, however, routinely showed the presence of these features.

• The test pit information led to an update to the site conceptual model (i.e., an expansion of the DNAPL source zone) which has obvious implications for future site remedial action. The soil boring sampling underestimated the extent of the source material and the quantity of free phase DNAPL present there. The test pit data supported an increased estimate of DNAPL mass and provided the basis for realistic remedial alternatives evaluations.

SUMMARY

The comparison of borehole and test pit information from the subject site has highlighted some of the advantages of test pits in characterizing DNAPL distribution and migration in complex geological settings. We recognize that in many DNAPL settings, test pitting will not be a practical investigative approach, either because of the depth to which the DNAPL has penetrated, subsurface conditions (e.g., the presence of bedrock), or site access constraints. However, the results from the subject site illustrate that where it is possible to use test pits to investigate the shallow portion of a DNAPL source zone, they can add substantially to the conceptualization of DNAPL distribution and migration processes and supplement soil boring information in a cost-efficient manner.

REFERENCES

Johnson, R. L. and B. Kueper. 1996. "Experimental Studies of the Movement of Chlorinated Solvent Compounds and other DNAPLs in the Vadose, Capillary, and Groundwater Zones." In J.F. Pankow and J.A. Cherry (Eds.), *Dense Chlorinated Solvents and other DNAPLs in Groundwater: History, Behavior, and Remediation*. Waterloo Press, Portland, OR.

Yu, S. 1995. "Transport and Fate of Chlorinated Hydrocarbons in the Vadose Zone – A Literature Review with Discussions on Regulatory Implications." *Journal of Soil Contamination*. 4(1): 25-56.

SURFACTANT-ENHANCED AQUIFER REMEDIATION OF PCE-DNAPL IN LOW-PERMEABILITY SAND

Frederick J. Holzmer (Duke Engineering & Services, Austin, Texas)
Gary A. Pope (The University of Texas at Austin, Texas)
Laura Yeh (NFESC, Port Hueneme, California)

ABSTRACT: A demonstration of surfactant-enhanced aquifer remediation (SEAR) was conducted during the spring of 1999 at Marine Corps Base, Camp Lejeune, NC. Surfactants were used to remediate a PCE-DNAPL zone located in a shallow aquifer beneath the Base dry cleaning facility at a depth of approximately 5-6 m below ground surface (bgs). The shallow aquifer is a relatively low-permeability formation composed of fine to very-fine sand, with a fining downward sequence in the bottom 0.6 m above a confining clay aquitard. A partitioning interwell tracer test (PITT) and soil core sampling were conducted before the SEAR to characterize the DNAPL zone. Soil cores were also taken following remediation to assess the performance of the SEAR. A total of 288 L of PCE was recovered during the surfactant flood and subsequent water flood through a combination of mobilization and enhanced solubilization. While the SEAR appeared ineffective in sweeping the low-permeability ($\sim 1 \times 10^{-4}$ cm/sec) clayey-silt zone overlaying the aquitard, data analysis of post-SEAR DNAPL conditions indicates that >92% of the source was removed from the upper, transmissive ($\sim 5 \times 10^{-4}$ cm/sec) zone. This demonstration indicates that the apparent practical lower limit for hydraulic conductivity for surfactant flooding at this time is approximately 1×10^{-4} cm/sec.

INTRODUCTION

A demonstration of surfactant-enhanced aquifer remediation (SEAR) was completed during 1999 at Site 88, the location of the central dry cleaning facility (Building 25) at Marine Corps Base (MCB) Camp Lejeune, NC. The demonstration included wastewater treatment of the SEAR effluent, as well as recovery and recycling of surfactant for reinjection during the surfactant flood. The primary purpose of the demonstration was to conduct additional field validation of SEAR for the remediation of sites contaminated with dense non-aqueous phase liquids (DNAPL). A secondary purpose was to evaluate the feasibility and cost benefit of surfactant regeneration and reuse during SEAR. This paper discusses only the subsurface remediation aspects of the project.

Site 88 is contaminated with immiscible phase tetrachloroethene (PCE), i.e., PCE DNAPL, which is the source of the Site 88 dissolved phase PCE ground-water plume. Soil sampling during initial DNAPL source zone investigations showed local DNAPL saturations in several locations to be >9%. Well installation and development revealed the presence of free-phase DNAPL in several locations of the test zone.

The DNAPL zone at Site 88 is located beneath Building 25, in the shallow surficial aquifer at a depth of approximately 16-20 ft (5-6 m), and includes an area that extends about 20 ft (6 m) north of the building. The DNAPL occurs immediately above and within a relatively low-permeability layer of silty sediments (hereafter referred to as the basal silt layer) that grade finer with depth from a sandy silt to a clayey silt until reaching a thick clay layer at about 20 ft bgs (6 m bgs). Characterization activities associated with the SEAR demonstration have revealed that this fining downward sequence can be roughly divided into three permeability zones: the upper zone (~16-17.5 ft bgs; 4.9-5.3 m bgs), the middle zone (~17.5-19 ft bgs; 5.3-5.8 m bgs), and the lower zone (~19-20 ft bgs; 5.8-6.1m bgs). The site conceptual model, or geosystem, is shown in cross section in Figure 1.

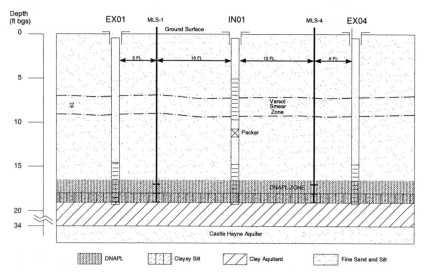

Figure 1. Generalized Geosystem Cross Section of DNAPL Zone at Site 88

The upper zone is generally characteristic of the shallow aquifer, which is primarily composed of fine to very-fine sand and is the most permeable of the three zones. The hydraulic conductivity of the upper zone is estimated to be about 5×10^{-4} cm/sec (1.4 ft/day). The hydraulic conductivity of the middle zone, which is predominantly composed of silt, is estimated to be approximately 1×10^{-4} cm/sec (0.28 ft/day), or about five times less permeable than the upper zone. The lower zone is predominantly composed of clayey silt, with a hydraulic conductivity that is believed to be approximately 5×10^{-5} cm/sec (0.14 ft/day) or perhaps even lower, although the permeability of the lower zone is not well characterized at this time. The upper- and middle-zone estimates of hydraulic conductivity are based on the analysis of pre-SEAR tracer test data from multi-level samplers. The conceptualization of decreasing permeability with depth in the DNAPL zone is supported by the results of grain-size analyses that were conducted on 72 soil samples from the bottom three feet of the test zone. The grain-size analyses confirm that the DNAPL zone is located above the clay aquitard in a fining downward sequence from fine sand to clayey silt.

SEAR DESIGN AND FIELD OPERATIONS

The Center for Petroleum and Geosystems Engineering at the University of Texas designed the SEAR demonstration. Due to the intended surfactant recycling operations, the design process included an extensive laboratory testing phase to select a surfactant with not only excellent subsurface performance, but also with characteristics amenable to recovering surfactant in the extracted effluent. As a result, a custom propoxylated alcohol ether sulfate surfactant, the Alfoterra 145-4-PO sulfate™, was developed and its performance was optimized with isopropanol as a cosolvent and calcium chloride as the electrolyte controlling phase behavior. The design process culminated with a series of simulations using UTCHEM (Delshad et al., 1996) to optimize the well configuration and flow rates for the SEAR demonstration. The SEAR design process for the Camp Lejeune SEAR demonstration is further discussed in Delshad et al. (2000).

The SEAR wellfield is situated in the portion of the DNAPL zone that lies just outside and north of Building 25, as shown in Figure 2. It consists of three central injection wells and six extraction wells arranged in a 3x3x3 divergent line-drive configuration. In addition, hydraulic control wells are located at each end of the centrally located row of injection wells. Thus, the test-zone wellfield comprises 11 wells in total. The test area formed by the 3x3x3 array of injection and extraction wells is 20 ft (6 m) wide by 30 ft (9 m) long. The wellfield configuration is shown in cross section in Figure 1. In preparation for the SEAR demonstration, the SEAR wellfield was used for free-phase DNAPL removal activities by water flooding as well as for additional aquifer characterization by conservative and partitioning interwell tracer tests.

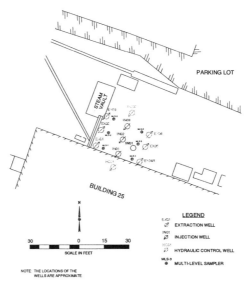

Figure 2. Demonstration Well Array and Multi-Level Sampling Points

The pre-SEAR partitioning interwell tracer test (PITT) was conducted during May/June 1998 to measure the volume and relative distribution of DNAPL present in the test zone before surfactant flooding. The results of this pre-SEAR PITT indicated that approximately 74-88 gallons (280-333 L) of DNAPL were present in the test zone (DE&S, 1999). Average DNAPL saturations were found to be highest in the portion of the test zone that is adjacent to Building 25, at about 4.5% saturation.

The SEAR demonstration at Site 88 included multiple phases of field activities from March to August 1999. Injection/extraction operations occurred continuously for 143 days, and included a pre-SEAR water flood, surfactant flood, post-SEAR water flood, post-SEAR PITT, and post-SEAR soil sampling.

The SEAR-injectate formulation consisted of 4% by weight (wt%) Alfoterra 145-4PO sulfate™ surfactant, 16 wt% isopropanol (IPA), and 0.16 - 0.19 wt% calcium chloride ($CaCl_2$) mixed with source water (i.e., site potable water). During the 58-day surfactant flood, 29,700 gallons (112,000 L) of the surfactant mixture was injected, which is equivalent to approximately five pore volumes injected into the test zone. The total mass injected was 9,718 pounds (lb) (4,410 kg) surfactant, 38,637 lb (16,620 kg) IPA and 427 lb (194 kg) $CaCl_2$.

Due to the low permeability of the DNAPL-contaminated sediments at Site 88, as well as the shallow conditions, injection and extraction flow rates during the SEAR demonstration were low relative to flow rates used at other surfactant flood projects. The total rate of surfactant injection was 0.4 gallons per minute (gpm) (1.5 L/min) and the total extraction rate was 1 gpm (3.8 L/min). Flow rates were varied during different phases of operations to improve the sweep of surfactant solution through the more highly-contaminated sections of the test zone.

RESULTS AND DISCUSSION

A total of 76 gallons (288 L) of PCE was recovered during the surfactant flood and subsequent water flood, of which approximately 32 gallons (121 L) of PCE were recovered as solubilized DNAPL and 44 gals (167 L) as free-phase DNAPL. In addition to enhancing the solubility of the DNAPL, the surfactant flood also enhanced the recovery of free-phase DNAPL as a result of lowering the interfacial tension (IFT) of the DNAPL. Lowering the IFT decreases the capillary forces retaining the DNAPL in the soil pores and thereby enhances its recovery by extraction wells. The DNAPL zone at Site 88 is protected below by an effective capillary barrier (i.e., thick clay aquitard) and the surfactant flood was designed and conducted with positive hydraulic control. Because of these site-specific circumstances and because of its greater mass removal efficiency, mobilization of DNAPL during the surfactant flood was desirable and intended by design.

Before the Site 88 SEAR demonstration, there was concern that injecting surfactants into such a fine-grained, low-permeability aquifer was not possible without causing plugging of the aquifer. However, laboratory testing during the design process eliminated the potential causes of surfactant-induced aquifer plugging, and the SEAR demonstration was completed with no measurable loss of permeability due to surfactant injection. This was a significant accomplishment of the laboratory portion of the demonstration project.

A post-SEAR PITT was conducted, along with soil core sampling, to measure the volume of DNAPL remaining in the test zone after the surfactant flood. The results of the post-SEAR PITT, however, proved to be unusable due to interference with a minor component of the surfactant that sorbed to the aquifer. Therefore, performance assessment of the SEAR demonstration was based on the results of the post-SEAR soil sampling. All soil samples collected for analysis of volatile organic compounds (VOCs) were field preserved with methanol.

The post-SEAR soil sample data was used to generate a 3-dimensional distribution of the DNAPL volume remaining in the test zone following the surfactant flood. The lateral distribution of DNAPL indicates that the majority of the DNAPL that remains in the test zone is located near the building, between wells EX01 and EX04. DNAPL volume decreases away from the building, in the area between wells EX02 and EX05, and very little DNAPL is present in the portion of the test zone that is farthest from the building, between wells EX03 and EX06. The vertical distribution of remaining DNAPL indicates that DNAPL was effectively removed from the more permeable sediments, generally above about 17.5 ft (5.3 m) bgs, and that DNAPL still remains in the lower permeability basal silt layer. These results are not unexpected, given that the highest pre-SEAR DNAPL saturations were near the building, as well as the expectation that it would be most difficult to remove DNAPL from the lowest permeability sediments at the site.

The multi-level sampler data for the IPA and surfactant response curves from the surfactant flood showed that little surfactant injectate penetrated (i.e. swept) the lower-permeability basal silt layer. It is considered unlikely that the tracers in the pre-SEAR PITT were able to sweep the low permeability basal silt zone much more effectively than the surfactants. Therefore, it is concluded that the pre-SEAR PITT did not measure much of the DNAPL that was present in the bottom 1-2 ft (0.3-0.6 m) of the aquifer. The pre-SEAR PITT did, however, accurately detect and measure the volume of DNAPL in the accessible (i.e. higher permeability) zone above approximately 18 ft (5.5 m) bgs.

Analysis of the post-SEAR soil core data indicates that approximately 5.2 ± 1.6 gallons (20 ± 6 L) of DNAPL remain in the zone that was effectively swept by the tracers and surfactant (i.e. the zone above approximately 18 ft bgs). In addition, data analysis from the post-SEAR soil cores indicates that approximately 23.5 ± 5.5 gallons (89 ± 21 L) remain in the mid-to-bottom zone that was not effectively penetrated by the tracers or surfactant (i.e. from 18 ft bgs down to the clay aquitard). The initial PITT estimated that the volume of DNAPL in the test zone before the surfactant flood was approximately 81 ± 7 gallons (307 ± 26 L). It is concluded here that the total volume of DNAPL present in the test zone before the surfactant flood is best represented by both the volume of DNAPL measured by the PITT plus the volume of DNAPL estimated (from soil core data analysis) for the zone below 18 ft (5.5 m) bgs, for a total pre-SEAR DNAPL volume of approximately 105 gallons (397 L).

Based on the volume of DNAPL distributed in the upper of the two permeability zones, it can be inferred that the surfactant flood recovered between 92% to 96% of the DNAPL that was present in the pore volume that was swept by the pre-SEAR PITT (i.e. above 18 ft bgs). Summing the two zones as a basis for the total pre-SEAR volume of DNAPL, the surfactant flood recovered approximately 72% of the DNAPL from the entire SEAR demonstration test zone, which includes all zones above the aquitard.

DNAPL recovery could have been increased from the middle (i.e., silt) zone by continuing the surfactant flood with additional pore volumes of surfactant and/or by increasing injection and extraction flow rates. There are certainly limits as to the magnitude and sustainability of higher flow rates in such a shallow, low-permeability aquifer without dewatering the zone of interest. However, there are well-development methods available that can be employed to increase the productivity of wells at Site 88. The use of polymer or foam during surfactant flooding may also be used to improve the recovery of DNAPL from lower-permeability zones. However, it is not known at this time if the permeability of the DNAPL zone at Site 88 is sufficient to support the use of polymer or foam.

CONCLUSIONS

The results of this demonstration indicate that the apparent practical lower limit for surfactant flooding at this time is approximately $1x10^{-4}$ cm/sec (0.28 ft/day). The surfactant flood effectively removed DNAPL from the upper, more permeable regions in the aquifer (above approximately 18 ft bgs) which is the zone where the highest pre-SEAR DNAPL saturations originally existed (based on pre-SEAR soil core saturations). Therefore, the greatest mass of DNAPL in the source zone at Site 88 occurred in the zone that was effectively remediated with surfactants, and the DNAPL that remains in the test zone following surfactant flooding is relatively isolated in the low-permeability basal silt layer. This is a very important result for the SEAR demonstration at Site 88. The resulting flux of dissolved PCE, from DNAPL remaining in the low-permeability silt layer, to the overlying, transmissive zone of the shallow aquifer will now be largely limited to diffusion rather than advection.

Finally, several observations can be made about the effects of permeability upon remediation costs. The duration of the Camp Lejeune SEAR demonstration was 143 days, which includes the post-SEAR PITT. When SEAR has been applied in higher permeability settings such as at Hill AFB, where the hydraulic conductivity is approximately two orders of magnitude greater (ranging from about $1x10^{-3}$ to $1x10^{-2}$ cm/sec (2.8 to 28 ft/day)), the duration for surfactant flooding and the accompanying tracer test has been about 1/10 as long as that required to treat a similar pore volume of aquifer at Camp Lejeune. Thus, lower permeability conditions will greatly extend the field time and thus the unit operating cost of SEAR, which is roughly proportional to the field time. However, it should be recognized that SEAR remediation is not unique in this respect -- low permeability aquifer conditions will increase the remediation time and therefore the implementation cost of all *in situ* DNAPL cleanup technologies.

ACKNOWLEDGEMENTS

The SEAR demonstration at Camp Lejeune was co-funded by the Environmental Security Technology Certification Program (ESTCP) and Naval Facilities Engineering Command, Atlantic Division (LANTDIV). We wish to express our gratitude to MCB Camp Lejeune for being the site host and for providing logistical support for the demonstration. The technical team involved a partnership between the Naval Facilities Engineering Service Center (NFESC), the US Environmental Protection Agency National Risk Management Research Laboratory (EPA NRMRL), the University of Texas (Austin), the University of Oklahoma (Norman) and Duke Engineering & Services. CONDEA Vista Company also provided generous support for the development of a custom surfactant. Duke Engineering & Services headed up field activities with assistance from Baker Environmental. Additional site support was provided by OHM Remediation Services Corporation.

REFERENCES

Delshad, M., G. A. Pope, and K. Sepehrnoori. 1996. "A Compositional Simulator for Modeling Surfactant-Enhanced Aquifer Remediation: 1. Formulation." *Journal of Contaminant Hydrology.* 23(4): 303-327.

Delshad, M., G. A. Pope, L. Yeh, and F. J. Holzmer. 2000. "Design of the Surfactant Flood at Camp Lejeune." In The Second International Conference on Remediation of Chlorinated and Recalcitrant Compounds, May 22-25, 2000 Monterey, California.

DE&S. 1999. *DNAPL Site Characterization Using a Partitioning Interwell Tracer Test at Site 88, Marine Corps Base, Camp Lejeune, North Carolina.* Report prepared for Naval Facilities Engineering Service Center, Port Hueneme, California.

ADVANCES IN THE FIELD IMPLEMENTATION OF PITTS AND SEAR

Hans W. Meinardus, Benito Casaus, Frederick J. Holzmer, Richard E. Jackson, Harold Linnemeyer, John T. Londergan, and Jeff A. K. Silva (Duke Engineering & Services, Austin Texas)

ABSTRACT: In the early 1990s, innovative NAPL characterization and removal technologies were developed from oil field predecessors. Surfactant-Enhanced Aquifer Remediation (SEAR) and the Partitioning Interwell Tracer Test (PITT) are complex flood-based technologies that require extremely reliable and robust process systems for successful and cost-effective field implementation. The first field applications of PITTs and SEAR consisted of pilot-scale demonstrations conducted in the mid-1990s. These efforts were labor intensive, requiring 24 hour manned operations several weeks long to control and monitor the batch-mode injection, extraction, and sampling operations manually. In 1998, the first full-scale PITTs and the application of SEAR at a low permeability site were made possible by the use of automated process systems. These systems are founded on a supervisory data acquisition and control system (SCADA) that monitors, controls, and logs process operations. Flow rates and in-line reagent mixing "on the fly" are controlled automatically. An automated in-line gas chromatograph (GC) system provides higher quality data in real time. The footprint of the process system has decreased markedly, and the units are now mobile. These advances have achieved a five-fold reduction in labor requirements, resulting in substantial reductions in the cost of these technologies.

INTRODUCTION

In June 1996, a team composed of DE&S (then known as INTERA, Inc.), UT Austin, and Radian conducted the first partitioning interwell tracer test or PITT (US Patent No, 5,905,036, US Patent No 6,003,365) and surfactant flood (Brown et al., 1997) at Hill Air Force Base's Operable Unit 2 (OU 2). To date, ten additional PITTs have been completed at this site since that first demonstration funded by the Air Force Center for Environmental Excellence (AFCEE). Four of these PITTs were conducted as performance assessments for two separate surfactant flood demonstrations (Hirasaki et al., 1997), five implemented at full-scale to quantitatively delineate the OU 2 DNAPL source zone (Meinardus et al. 1999), and one completed to access the performance of a full-scale surfactant enhanced aquifer remediation (SEAR). Over this four-year period of PITT and SEAR operations at OU 2, a number of innovations and improvements have significantly enhanced these technologies and drastically reduced the cost of implementing them. This paper chronicles the history of some of the more important of these improvements and outlines the current state of the art in the field implementation of flood-based technologies.

PROCESS SYSTEM AUTOMATION

One of the most visible changes in PITT and SEAR field operations over the last four years has been a reduction in the manpower required to conduct these tests. The labor requirements for the various PITTs and SEAR conducted at OU2 are listed chronologically on Table 1. The decrease in the amount of field personnel evidenced in this table are largely attributable to the automation of time consuming and labor intensive field tasks.

TABLE 1. Decreases in PITT/SEAR Field Labor Requirements

Date	Project	Labor Requirements
June 1996	AFCEE SEAR -Phase I	7 days-24 hrs: 3 shifts per day of 3 people + site manager + 2 chemists in onsite lab = 12+ person field staff for approximately 28 days.
August 1996	AFCEE SEAR Phase II	7 days-24 hrs: 3 shifts of 2 people + site manager + 2 chemists in onsite lab = 9 person field staff for approximately X days.
June 1997	AATDF Surfactant/Foam Process	7 days-24 hrs: 3 shifts of 2 people + site manager = field staff of 7 people.
June 1998 – Nov 1999	OU2 Source Zone Delineation	7 days-8 to10 hrs: – 1 shift of 3 to 4 people at start of a large scale PITT, reduced to two person staff after peak concentration recovered. Average time for a PITT = 22 days
Jan 2000	Full-Scale SEAR	7 days-8 to 10 hrs: - 1 shift of 4 to 5 people during first part of surfactant injection, 3 person field crew during post-injection brine flood, 2 person crew during following water flood for 40 days.

Conducting a field-scale PITT or SEAR requires an injection system, an extraction system, a sampling system, and a fluid monitoring and control system. Collectively, these components make up the PITT/SEAR process system, an example of which is shown in Figure 1. Early pilot-scale tests utilized static systems constructed in place. In order to decrease the mobilization and setup time, and to facilitate the rapid deployment of consecutive PITTs, the process system became modular and portable in design. One of the most visible changes in the field implementation of PITTs and SEARs over the last four years has been the decrease in the footprint required for process systems. A comparison of the 1996 and 1998 process system plans is shown on Figure 2. The secondary containment system used in 1996 covered 4800 square feet. In 1998, the process system for a PITT covered approximately 600 square feet.

Furthermore, once a PITT begins, the process system must continue to function without major interruptions until practically the entire tracer mass is recovered at the extraction wells. Prior to the first of the full-scale PITTs, this requirement was met by manning the test continuously, making the PITT a very labor-intensive field effort. However, the lengthy duration of each of the large-scale OU2 PITTs necessitated that the process system be dependably automated to make the tests cost effective.

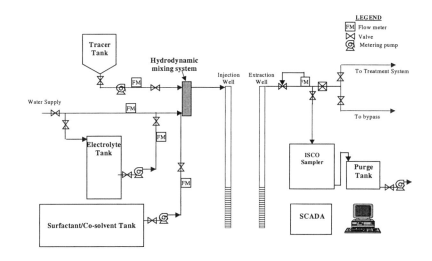

FIGURE 1. 1998 PITT/SEAR Process System Schematic

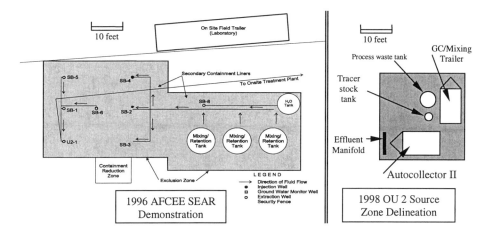

FIGURE 2. Comparison of 1996 and 1998 PITT/SEAR Process System Plan

Fluid Mixing. Accurate and efficient mixing of injectate is crucial to the technical success of a flood-based technology such as a PITT or SEAR. During the 1996 AFCEE SEAR and 1997 AATDF Surfactant/Foam Flood, injectate was mixed in a batch mode. Mixing batches required the use of recirculation and injection pumps. Attaining a homogenous tracer batch typically took 24 to 48 hours of mixing, while surfactant/cosolvent batches typically took 12 hours. Moreover, the surfactant and co-solvent was staged on site in 200 L drums, while each of the acohols in the tracer suite were purchased and stored in small individual containers. Flow control during these demonstrations was manually monitored, recorded and controlled using flow meters and needle valves. Periodic bucket testing provided verification of the flow meter readings.

In June 1998, in-line mixing "on the fly" was introduced during the first full-scale PITT. The three 3000 gal batch tanks, recirculation pumps, and injection pumps were replaced with a conical bottomed 200 gal tracer stock tank. For the full-scale PITTs, the tracer slugs were mixed automatically and injected extremely accurately directly into the injection line from the tracer stock tank with chemical metering pumps. A hydrodynamic mixing vessel was used to ensure complete mixing of the tracers as they were being injected. Finally, the tracer suite of five alcohols was blended in the correct proportions by the vendor and shipped to the site in 200 L drums.

In January of 2000, the first full-scale SEAR was implemented at OU 2. An upgraded version of the PITT process system was used to conduct the remedial operation. Similar to the approach used for the PITT tracer suites, the surfactant was blended with the co-solvent at the manufacturing plant and shipped to the site in bulk. Concentrated electrolyte batches were mixed in large tanks. During the surfactant injection period, the surfactant solution and electrolyte was metered into the injectate line using chemical metering pumps.

Flow Rates. Closely related to the requirement for accurate injectate mixing is the need to control and monitor injection and extraction flow rates. During a PITT, the injection and extraction flow rates need to be held constant for the duration of the test. In addition, an accurate record of each test's actual flow rates are required to correctly analyze the tracer test results (Jin et al., 1995). Flow rates during the 1996 AFCEE SEAR and 1997 AATDF Surfactant/Foam Flood demonstrations were manually monitored, recorded and controlled using flow meters and needle valves. Periodic bucket testing provided verification of the flow meter readings.

The full-scale PITT and SEAR process system relied on an automated flow control system. The flow rates for each injection and extraction well were monitored continuously using inline flow meters. Based on the flow-meter reading, an actuated valve on each line adjusted the flow automatically to ensure that it was within 10% of the desired flow rate. The desired flow rate was set in the control system (see below) which controlled the entire system and logged the flow rates electronically every 15 minutes. In addition to the flow meter and actuated valve, each line was equipped with a check valve, a manual flow control valve and a bypass so that repairs to the line could be made without interrupting the test.

Sampling and Data Acquisition. The 1996 AFCEE SEAR demonstration relied on manual sample collection at spigots on a sampling panel located in the well field. Handwritten sample control logs, sample labels, and chain of custody forms were used to manage over 1500 aqueous samples. Samples were analyzed at a laboratory on site. A simple data logger and pressure transducers were used to collect water levels, and conductivity and pH measurements were collected manually. In June 1997, samples for the AATDF demonstration were collected automatically with a system of sample loops and actuated rotary valves mounted in a trailer (Autocollector I). Manual purging of the sample loops into glass vials was required to complete the sampling process. Samples were shipped to a laboratory offsite for analysis. Sample logs were generated electronically, and the sample labels were preprinted. Conductivity and pH were monitored automatically during the surfactant/foam using a HYDROLAB data sonde. A more sophisticated datalogger with real time data display capability was used to monitor and log water levels, gas pressures, and tank temperatures.

The full-scale PITTs started in June 1998 utilized refrigerated ISCO™ VOC samplers to collect 40 ml samples automatically. The filled glass vials were packaged and shipped to an offsite laboratory. Sample logs, labels, and chain of custody forms were generated automatically using a database and the electronic logs generated by a supervisory control and data acquisition system (SCADA).

In July 1998, a significant advancement in the sampling process was achieved with a prototype in-line gas chromatograph system. This system utilized an automated stream selector to provide real time analytical results from two central extraction wells that were sampled automatically approximately every 80 minutes. Small diameter sample lines diverted a fraction of the fluids produced from the northern and southern central extraction wells to an automated multiport valve programmed to select between the wells and purge the system after each sample run. Micro-liter samples from each well were injected into the GC and analyzed at regular intervals using a flow-through auto-injection system. The information obtained from the in-line GC proved useful in tracking the progress of each PITT. The on-line GC provided excellent data quality as compared to conventional sample collection. Based on the success of the prototype, the in-line GC was upgraded and became the primary sampling system for the fifth full-scale PITT and the first full-scale SEAR conducted at OU2. Thereafter the ISCO VOC samplers became a backup and confirmatory sampling system.

THE PITT/SEAR PROCESS SYSTEM

The state-of-the-art process system used to conduct the four large-scale PITTs is the culmination of these improvements. The majority of this system is mounted within two insulated trailers to provide the desired mobility and to protect the equipment from the weather. One trailer houses the tracer mixing system and an on-line GC used to obtain tracer analyses in real time from the central extraction wells. The sampling system, flow monitoring and control system, and the process control system are located in the other trailer, known as the Autocollector II. Connections between the trailers, the well array being tested, the water source, and the treatment plant are made with flexible hoses to allow the layout to be adapted to the available space, and to facilitate moving the system quickly. The process system has a remote alarm, access and control capability so that it can be left unattended at night. In June 1998, during the first full-scale PITT, this alarm and automated paging capability, coupled with remote system access and control via modem and laptop, allowed unmanned PITT/SEAR operations for the first time. The resulting reduction in labor requirements has been one of the most significant advances in the implementation of SEAR and PITTs. The following sections describe the current process system in use at OU2.

The Supervisory Control and Data Acquisition (SCADA) System. One of the most important components of the automated PITT process system, the SCADA electronically monitors and controls injection and extraction flow rates, the tracer injection system, and the automated sampling system. In addition to controlling the sampling system, the SCADA electronically maintains all sample acquisition data in the form of an electronic sampling log. The SCADA also logs and stores all of the system parameters it monitors, including the injection and extraction flow rates, effluent conductivity, and aquifer water-levels in the well array via pressure transducers installed in the injection, extraction, and monitor wells. The resulting data is displayed in real time on a virtual instrument panel, and can also be displayed graphically in the form of trend plots. In the event of a power or system failure, or if a monitored parameter falls out of a set range, the system uses its modem to page the operator with an alarm code identifying the type of problem. The SCADA system can be accessed and controlled remotely via a PC modem. Finally, an inline conductivity probe monitored by the SCADA is also mounted on the two central extraction well lines so that the progress of the tracer test can be tracked with conductivity measurements.

The Sampling System. The accuracy of a PITT largely depends on the quality of the tracer breakthrough data obtained from the samples taken during the test (see the error analysis in Section 6.2). Moreover, enough samples need to be collected from each extraction and monitor well to adequately characterize the tracer curves, necessitating that a large number of samples be collected in a relatively short amount of time. The PITT process system uses an automated in-line GC sampling system mounted in a trailer to provide quality samples without the labor burden of manual sampling. The on-line GC apparatus consists of a stream selector and a GC equipped with a modular liquid auto-injector. The stream selector governs fluid flow from each well location, and directs flow through the auto-injector. Both the auto-injector and the GC are programmed to operate synchronously. A personal computer is used to operate the chromatography softwareand to maintain a record of tracer breakthrough at each well location, sampling schedule(s), GC calibration records, and GC-related QA/QC.

The two monitor wells in each well array provide data from between the central injection well and the central extraction wells without significantly impacting the PITT flow field. This is achieved by installing a small pneumatic double-valve pump in each well and pumping them continuously at approximately 0.03 gpm (120 mL/min). Previous experience has shown that a continuously pumped monitor well provides data of much better quality than an intermittently pumped one. The pumps were connected to the Autocollector sampling system and sampled in the same fashion as the extraction wells.

The effluent produced by the monitor wells is collected in a wastewater tank, along with the purge water from the Autocollector. This process wastewater tank is periodically pumped into the effluent line to the SRS for treatment. With all of these component systems in place, fieldwork primarily consists of monitoring, operating, and maintaining the PITT process system.

CONCLUSIONS

Significant advances in implementing PITTs and SEAR have been made since the first technology demonstrations in 1996. Improvements have predominantly involved automating the most labor intensive field tasks and thereby alleviating the need to man the process system around the clock. The result has been a drastic five-fold reduction in labor costs for these types of projects. Future improvements may include adding the electrolyte to the bulk surfactant/co-solvent mixture at the manufacturing plant, developing the capability to inject tracers through temporary drive points installed by direct push methods, and enhancing the electronic transfer of data from the SCADA to the sampling data base and the in-line GC system. Advances and accompanying cost efficiencies will continue to be realized as these technologies are applied at an increasing number of DNAPL and LNAPL sites.

REFERENCES

Brown, C.L., M. Delshad, V. Dwarakanath, R. E. Jackson, J. T. Londergan, H. W. Meinardus, D. C. McKinney, T. Oolman, G. A. Pope, and W. H. Wade. 1999. "Demonstration of Surfactant Flooding of an Alluvial Aquifer Contaminated with DNAPL." In *Innovative Subsurface Remediation*, ACS Symposium Series #725, American Chemical Society, Washington DC.

Hirasaki, G. H., C. A. Miller, R. Szafranski, D. Tanzil, J. B. Lawson, H. W. Meinardus, M. Jin, J. T. Londergan, R. E. Jackson, G. A. Pope, and W. H. Wade. 1997. "Field Demonstration of the Surfactant/Foam Process for Aquifer Remediation." *Society of Petroleum Engineers, Inc. Annual Technical Conference and Exhibition*, San Antonio, Texas Oct 5-8. #39292

Jin, M., M. Delshad, V. Dwarakanath, D. C. McKinney, G. A. Pope, K. Sepehrnoori, C. E. Tilburg, and R. E. Jackson. 1995. "Partitioning Tracer Test For Detection, Estimation And Remediation Performance Assessment Of Subsurface Nonaqueous Phase Liquids." *Water Resources Research.* 31(5): 1201-1211.

Meinardus, H. W., V. Dwarakanath, M. Fort, R. E. Jackson, M. Jin, J. S. Ginn, and G. C. Stotler. 1999. "The Delineation of a DNAPL Source Zone with Partitioning Interwell Tracer Tests." In *Proceedings of the 2000 Petroleum Hydrocarbons and Organic Chemicals in Ground Water: Prevention, Detection, and Remediation, Conference and Exposition*, Nov 17-19 Houston TX

United States Patent 5,905,036. Characterization of Organic Contaminants and Assessment of Remediation Performance in Subsurface Formations. Inventors Gary Pope and Richard E. Jackson, May 18th 1999.

United States Patent 5,905,036. Characterization of Organic Contaminants and Assessment of Remediation Performance in Subsurface Formations. Inventors Gary Pope and Richard E. Jackson, Dec 21, 1999

DESIGN OF THE SURFACTANT FLOOD AT CAMP LEJEUNE

Mojdeh Delshad and Gary A. Pope (The University of Texas at Austin, Austin, Texas)
Laura Yeh (Naval Facilities Engineering Service Center, Port Hueneme, California)
Frederick J. Holzmer (Duke Engineering and Services, Austin, Texas)

ABSTRACT: A surfactant flood to remove perchloroethylene (PCE) from the groundwater below a dry cleaning facility at Marine Corps Base Camp Lejeune in North Carolina was conducted during the summer of 1999. The design and operation of both the subsurface and surface treatment processes was a large effort involving people from several organizations. Here we present only the design aspects of the subsurface surfactant flood with the emphasis on the flow and transport modeling. The University of Texas numerical model UTCHEM was used for this purpose. The very low hydraulic conductivity in the DNAPL zone at the bottom of this shallow aquifer made this a very challenging project design. Numerous simulations were performed to determine test design variables. These include wellfield configuration and rates, duration of injection and composition of surfactant solution, and hydraulic control achievement using hydraulic control wells. We discuss the simulation results performed prior to the demonstration and emphasize the sensitivity of the model predictions to most important but often uncertain design parameters such as permeability and heterogeneity and initial PCE-DNAPL volume and its distribution.

INTRODUCTION

A surfactant-enhanced aquifer remediation (SEAR) demonstration to remove PCE from the groundwater at site 88, below a dry cleaning facility at Marine Corps Base Camp Lejeune in North Carolina, was completed during 1999. The objectives of this demonstration were (1) further validation of SEAR for dense nonaqueous phase liquid (DNAPL) removal and (2) evaluation of feasibility and cost benefits of surfactant regeneration and reuse during SEAR. A total of 288 L of PCE was recovered during the surfactant test. This corresponds to a recovery of about 92% from the upper permeable zone where most of the contaminant source was initially located. Details on the site description, field-test operations, and results are given in Holzmer *et al.* (2000).

The design of the SEAR demonstration was completed after integrating the information obtained during the site characterization activities and the laboratory studies into the numerical model. The site characterization activities included the pre-SEAR tracer tests. The results of the field tracer tests were used to modify the geosystem model for the final design of the surfactant flood. Subsequently, the most important SEAR design variables were identified by selectively changing model parameters and determining effect on the simulation results.

The University of Texas flow and transport numerical simulator called UTCHEM was used for all geosystem model development (Delshad *et al.,* 1996). UTCHEM is a three-dimensional, multiphase, multicomponent chemical compositional simulator capable of modeling DNAPL migration and groundwater flow and transport in aquifers. Numerical simulation makes possible the study of parameters critical to the design of surfactant tests. These include contaminant solubilization, free-phase mobilization, organic and surfactant adsorption, interfacial tension reduction, capillary desaturation, dispersion/diffusion, and the complex phase behavior of groundwater, surfactant, NAPL, cosolvent, and electrolyte mixtures. We used the model not only to predict the performance but to address issues critical to gaining approval to conduct the demonstration. These issues included: (1) demonstration of hydraulic containment, (2) prediction of recoveries of injected chemicals, (3) prediction of DNAPL recovery, and (4) prediction of the final concentrations of injected chemicals and source contaminant.

Numerous simulations were conducted to develop the recommended design for surfactant flooding at Camp Lejeune. The objectives of the simulation study were to:

- Determine the time required for each test segment: pre-surfactant water injection, surfactant injection, and post-surfactant water injection
- Determine mass of surfactant, alcohol, and electrolytes recommended for each segment of the test
- Determine test design parameters such as number of hydraulic control, injection, and extraction wells, well locations, and well rates.
- Estimate effluent concentrations of contaminant, surfactant, alcohol, and electrolytes during and at the end of the test
- Estimate mass of contaminants, surfactant, alcohol, and electrolytes that remain in the volume of the test zone at the end of the test
- Evaluate the sensitivity of the performance of the proposed design to critical aquifer properties such as permeability and degree of heterogeneity, process parameters such as microemulsion viscosity, and operational parameters such as flow interruption in the middle of the surfactant injection

DESIGN OF THE SEAR FIELD TEST

Site Description and Characterization.   The site characterization included the following: aquifer stratigraphy and aquiclude topography, porosity and permeability distribution, soil, groundwater, and contaminant constituents and distribution, hydraulic gradient direction and magnitude, aquifer temperature and pH. The field site characterization conduced by Duke Engineering & Services (DE&S, 1999a) was based on the following site data: soil borings, well logs, water levels, soil contaminant measurements, DNAPL and groundwater sampling and analysis, hydraulic testing, and historical pumping data. The DNAPL zone was found to be about 5-6 m below ground surface. The aquifer soil has a relatively low permeability with about an order of magnitude smaller permeability in the bottom 0.3 m of the aquifer. Because of the low aquifer permeability and the limited thickness, each pore volume of the surfactant required about 12 days to inject.

Prior to the installation of the SEAR wellfield, several well patterns with different number of wells were simulated. The most efficient well pattern based on site hydrogeological data was a line of 3 central injection wells and 6 extraction wells arranged in a divergent line-drive pattern as shown in Figure 1. To maintain hydraulic control and to ensure adequate sweep efficiency in the wellfield, each injection and extraction well was spaced about 3 m apart and the distance between any pair of injection and extraction wells was about 4.6 m. A dual injection system was used in the three injectors to prevent upward migration of injectate and to focus the flow paths of injected surfactant through the DNAPL zone along the bottom portion of the aquifer. Water was injected in the upper screen with simultaneous injection of surfactant mixture in the lower screen. Similarly, to provide further hydraulic control of the injected fluids, several scenarios to identify the number of hydraulic control wells and their locations were simulated. Two hydraulic control wells located on each end of the line of injectors (Figure 1) were found to be adequate to achieve hydraulic control of the test zone.

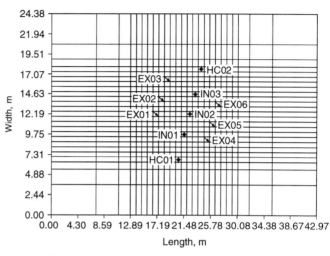

FIGURE 1. Simulation grid and well locations.

The results of the pre-SEAR partitioning interwell tracer test conducted during May/June 1998 indicated that approximately 280 to 333 L of DNAPL were present in the test zone. The highest average DNAPL saturations found were in the range of 4.5%. The spatial distribution of DNAPL saturation was highly variable in both areal and vertical directions.

Surfactant Phase Behavior. In UTCHEM, phase behavior parameters define the solubility of the organic contaminant in the microemulsion phase as a function of surfactant, cosolvent, and electrolyte concentrations. Solubility data for Camp Lejeune DNAPL obtained from laboratory experiments were used to calibrate the phase behavior model parameters (Ooi, 1998). The DNAPL is ~ 99% PCE. In addition to modeling surfactant phase behavior, the physical property model parameters such as microemulsion viscosity and microemulsion-DNAPL interfacial tension were also calibrated against experimental measurements.

A total of 155 surfactant formulations were screened by observing the phase behavior and measuring selected phase properties such as microemulsion viscosity until an optimum mixture was found (Weerasooriya *et al.*, 2000). The target properties of the optimum mixture include (1) high DNAPLs solubilization, (2) fast coalescence to a microemulsion (less than a day), (3) low microemulsion viscosity, and (4) acceptable ultrafiltration characteristics. The surfactant composition used in the field test was a mixture of 4 wt% Alfoterra^{TM} 145-4-PO sodium ether sulfate, 16 wt% isopropyl alcohol (IPA), and 0.16-0.19 wt% calcium chloride mixed with the source water. The alfoterra is made from a branched alcohol with 14 to 15 carbon atoms by propoxylating and then sulfating the alcohol. The contaminant solubilization was about 300,000 mg/L at 0.16 wt% $CaCl_2$ and about 700,000 mg/L at 0.19 wt% $CaCl_2$.

Aquifer Model Development and Simulations. The plan view of the three-dimensional grid used for the design of the Camp Lejeune surfactant flood is shown in Figure 1 and described in Table 1. A total of 10,000 gridblocks using a 25x25x16 mesh 43 m long and 24.4 m wide was used. The nearly 4 m saturated thickness of the aquifer was divided into 16 nonuniform numerical layers vertically. The top elevation of the numerical grid corresponds to about 5.5 m amsl. Both layered and stochastic permeability distributions were used for the design simulations. The aquitard gridblocks were identified based on the clay elevation data mapped to the grid and were assigned a very low porosity and permeability. A ratio of horizontal to vertical permeability of 0.1 was used. Design simulations were conducted with permeability fields with different values of average and variance and initial DNAPL volume and saturation distributions.

RESULTS AND DISCUSSIONS

We conducted numerous simulations to design the SEAR test for the Camp Lejeune. Here we only discuss the results of the two predictive simulations conducted prior to the field test to illustrate the sensitivity of variations in the permeability on the contaminant recovery. Simulation ISA7m assumed a lower permeability for the bottom 0.6 m of the aquifer, whereas simulation ISA26m assumed a higher permeability. We also discuss the results of our brief attempt to history match the field data.

Simulation No. ISA7m. The injection strategy given in Table 2 included 2 days of water preflush followed by 30 days of surfactant solution. A summary of flow rates for various sections of the flood is given in Table 3. The hydraulic conductivity was 4×10^{-4} cm/s for the top 12 layers (3.3 m) and 8×10^{-5} cm/s for the bottom 4 layers (0.6 m) i.e., a permeability contrast of 5:1. The initial DNAPL saturation increased with depth for the bottom 0.6 m of the aquifer. This corresponds to an average DNAPL saturation of 0.02 within the wellfield. A comparison between the predicted and measured dissolved PCE concentration at extraction well EX01 is shown in Figure 2 and agree very well. The peak PCE concentration predicted from the model is 1500 mg/L compared to the field-observed value of 2800 mg/L. The total simulated PCE recovery (dissolved and free-phase) from all the wells at the end of the flood (100 days) was 310 L compared to a recovery of 288 L

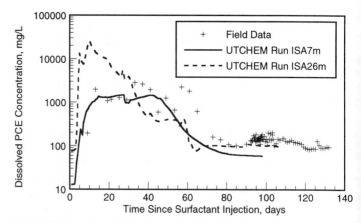

FIGURE 2. Comparison of predicted dissolved contaminant concentration and measured concentration in extraction well EX01.

measured in the field demonstration.

Simulation No. ISA26m. The results from simulation ISA26m were reported in the SEAR Work Plan (DE&S, 1999b) and were used as the final basis for the design and operation of the surfactant flood at Camp Lejeune. The aquifer permeability was modeled using a random correlated permeability field with an average hydraulic conductivity of 4×10^{-4} cm/s with a standard deviation of log k of 1. The hydraulic conductivity of the bottom 0.3 m was then reduced by a factor of 4 to an average of 10^{-4} cm/s. The injection strategy given in Table 2 included 6 days of electrolyte preflush with 0.22 wt% $CaCl_2$ followed by 48 days of surfactant solution. Injection and extraction rates during the preflush and surfactant injection were reduced compared to those used during the field tracer test or postwater flush because of the higher viscosity of surfactant solution (2.5 mPa.s) compared to the water. A summary of flow rates for various sections of the flood is given in Table 3. The reduction in the rates will prolong the surfactant injection test, however, it reduces the risk of excessive water buildup near the injection wells or excessive drawdown near the extraction wells. This was especially critical for this shallow aquifer. A comparison of the predicted and measured PCE concentration at extraction well EX01 is shown in Figure 2. The peak PCE concentration in simulation ISA26m was 25,000 mg/L whereas that observed in the field was an order of magnitude smaller of about 2800 mg/L. The predicted surfactant and IPA

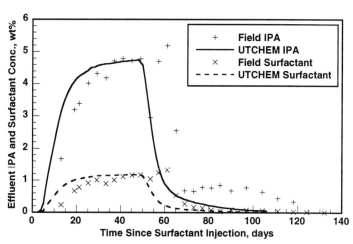

FIGURE 3. Comparison of field and predicted surfactant and IPA concentration at well EX01 for run ISA26m.

effluent concentrations in well EX01 are compared with the measured data in Figure 3. The agreement is not as good in the other extraction wells due to highly heterogeneous nature of both permeability and DNAPL saturation distributions not accurately accounted for in the model.

Discussion of UTCHEM Predictive Simulations. From Figure 2, it is evident that the match between predicted and measured PCE concentrations for run ISA7m is better than for run ISA26m. This is because in run ISA7m, the hydraulic conductivity of the bottom 0.6 m was 8×10^{-5} cm/s which is 5 times lower than that in the upper 4.3 m. In comparison to ISA26m, the hydraulic conductivity of the bottom 0.3 m was 10^{-4} cm/s, which is 4 times lower than that in the upper 4.6 m. The combination of a thicker and less permeable bottom 0.6 m in ISA7m compared to a thinner and more permeable bottom 0.30 m in ISA26m explains overestimate of the effluent PCE solubilities. Based on these results, it was inferred that the permeability contrast between the less permeable bottom of the aquifer and the other zones is at least a factor of 5. This can also explain a partial remediation of the lower permeability bottom zone as was observed during the field demonstration (Holzmer *et al.*, 2000).

Field History Match Simulation. A preliminary effort was made to qualitatively match the results of the field test. Adjustment in the UTCHEM input included the well rates and duration of the initial water flush and surfactant injection that were slightly altered during the field test compared to those of the final design simulation ISA26m. The injection and extraction rates and strategies are summarized in Tables 3 and 4. The rates were altered from the design rates to improve the sweep efficiency of the surfactant solution through the more highly contaminated sections of the test zone. Other adjustments in the model were in the spatial distribution of the DNAPL and permeability and its variation with depth. An attempt was made to approximate the grading of the DNAPL saturation across the wellfield, but the actual variations are more complex. The DNAPL volume is about 265 L within a pore volume of about 22,474 L in this simulation. This is an average DNAPL saturation of about 0.0118 within a swept

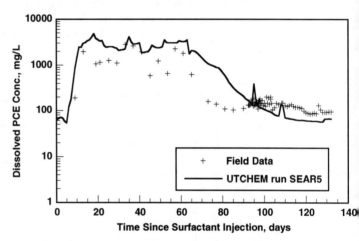

FIGURE 4. Comparison of measured and history match of dissolved PCE concentration in extraction well EX01.

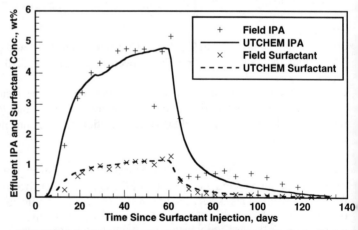

FIGURE 5. Comparison of measured and history match of surfactant and IPA concentrations in extraction well EX01 for simulation SEAR5.

pore volume similar to that estimated from the pre-SEAR tracer test. The permeability was modeled as three geological layers as given in Table 4. A permeability contrast of 20 was used between the upper high permeability zone and the bottom zone right above the clay aquitard. The effluent PCE concentrations compare favorably to those measured at the field in most of the extraction wells. Comparison of measured and history match of dissolved PCE concentrations for extraction well EX01 is shown in Figure 4. The final DNAPL volume within a swept volume of 22,474 L was 76.5 L corresponding to an average DNAPL saturation of 0.0034. This gives a PCE recovery of about 72% for run SEAR5. The simulated surfactant and IPA concentrations and subsequently the recoveries were higher than the field data in most of the extraction wells. The breakthrough times and the peak concentrations of both IPA and surfactant were closely matched in well EX01 as shown in Figure 5. Plausible explanations for the lower field surfactant recoveries are (1) higher surfactant adsorption in the bottom of the aquifer with high clay content and (2) biodegradation of surfactant, and (3) fluctuations in the injected surfactant concentration compared to the design value of 4 wt%.

SUMMARY AND CONCLUSIONS

The surfactant flood for the Camp Lejeune PCE-DNAPL site was simulated using UTCHEM simulator. The design was based on the available surfactant phase behavior data and geosystem model calibrated against the pre-SEAR tracer test. The results of model predictions provided critical guidelines for the field operation. These include the wellfield design, hydraulic containment, well rates, the frequency of the sampling for effluent analysis, and effluent concentrations necessary for surface treatment and surfactant recycling operations.

ACKNOWLEDGMENT

The SEAR demonstration at Camp Lejeune was co-funded by the Environmental Security Technology Certification Program (ESTCP) and LANTDIV. CONDEA Vista Company provided generous support in the development of a custom alfoterra surfactant.

REFERENCES

Delshad, M., G. A. Pope, and K. Sepehrnoori. 1996. "A Compositional Simulator for Modeling Surfactant-Enhanced Aquifer Remediation: 1. Formulation." *Journal of Contaminant Hydrology*. 23(4): 303-327.

DE&S. 1999a. *DNAPL Site Characterization Using a Partitioning Interwell Tracer Test at Site 88, Marine Corps Base, Camp Lejeune, North Carolina*. Report prepared for Naval Facilities Engineering Service Center, Port Hueneme, California.

DE&S. 1999b. *Work Plan for Surfactant-Enhanced Aquifer Remediation Demonstration, Site 88 Marine Corps Base, Camp Lejeune, North Carolina*. Report prepared for Naval Facilities Engineering Service Center, Port Hueneme, California.

Holzmer, F.J., G.A. Pope, and L. Yeh. 2000. "Surfactant-Enhanced Aquifer Remediation of PCE-DNAPL in Low-Permeability Sand." In The Second International Conference on Remediation of Chlorinated and Recalcitrant Compounds, May 22-25, Monterey, California.

Ooi, K.C. 1998. "Laboratory Evaluation of Surfactant Remediation of Nonaqueous Phase Liquids." M.S. Thesis, The University of Texas at Austin.

Weerasooriya, V., S.L. Yeh, and G.A. Pope. 2000. "Integrated Demonstration of Surfactant-Enhanced Aquifer Remediation (SEAR) with Surfactant Regeneration and Reuse." *Surfactant-Based Separations: Science and Technology*, J.F. Scamehorn and J.H. Harwell (eds.), ACS Symposium Series 740, American Chemical Society, Washington, DC.

TABLE 1. Grid and aquifer properties for SEAR design.

Dimension	42.97 x 24.38 x 3.96 meters
Mesh	xyz: 25x25x16 (10,000 gridblocks)
Porosity	0.28
Aquifer pore volume	937,146 liters
Smallest gridblock size	0.9144 x 0.6096 x 0.1524 meters
Largest gridblock size	7.312 x 3.657 x 0.6096 meters
Boundary conditions	Impervious top, bottom, north, and south; constant potential west-east boundaries with a hydraulic gradient of 0.0123 m/m
Initial pressure	Atmospheric pressure in top layer, hydrostatic distribution in vertical
Initial electrolyte concentration	0.1 wt% calcium chloride

Table 2. Injection strategies used in the SEAR simulations.

	Run ISA7m		Run ISA26m		Run SEAR5	
Process	days	rate	days	rate	days	rate
Water preflush	2	No. A	6	No. B	8	No. B
Surfactant flood	30	No. A	48	No. B	58	No. C
Water postflush	68	No. A	58	No. A	74	No. A

Table 3. Well rates (m³/day) used in SEAR simulations.

Wells	Rate No. A	Rate No. B	Rate No. C
Upper screen injection: IN1, IN2, IN3	0.436	0.436	0.436
Injection: IN01, IN02, IN03	1.09	0.727	0.545 - 0.927
Hydraulic control: HC01, HC02	1.635	1.09	1.09
Extraction: EX01-EX06	1.362	0.908	0.709 to 1.2

Table 4. Hydraulic conductivity used in the history match simulation SEAR5.

Simulation Layer number	Thickness, m	Hydraulic conductivity, cm/s
1-12	3.35	2×10^{-4}
13-14	0.30	5×10^{-5}
15-16	0.30	10^{-5}

EVALUATION OF SURFACTANT FORMULATIONS FOR TREATMENT OF A PCE-CONTAMINATED FIELD SITE

C. Andrew Ramsburg and Kurt D. Pennell
Georgia Institute of Technology, Atlanta, Georgia, USA

ABSTRACT: Laboratory studies and economic analyses were performed to select surfactant formulations for pilot-scale treatment of a tetrachloroethylene (PCE)-contaminated field site located in Oscoda, Michigan. Following a preliminary screening, detailed laboratory treatability tests, consisting of batch and 2-D cell experiments, were conducted to evaluate two surfactant formulations; 4% Tween 80 + 500 mg/L $CaCl_2$ and 8% Aerosol MA/IPA + 15,000 mg/L NaCl + 1000 mg/L $CaCl_2$. In the 2-D aquifer cell studies, approximately 8 pore volumes of 4% Tween 80 solution were required to recover 53% of the PCE initially released. Approximately 3 pore volumes of 8% Aerosol MA/IPA solution recovered 78% of the PCE released. Mobilization of PCE free product was observed with 8% MA/IPA, but not for the 4% Tween 80, consistent with total trapping number calculations. Pilot-scale costs for remediation using the 4% Tween 80 solution were estimated to be \$212,000, while the costs for the 8% MA/IPA solution were estimated to be \$239,000. In comparison, pilot-scale pump and treat costs were estimated to be \$307,000.

INTRODUCTION

Remediation of dense nonaqueous phase liquid (DNAPL) source zones is problematic due to the relatively low aqueous solubility of DNAPLs and their complex distribution within the subsurface. The most commonly implemented technology for the remediation of DNAPL-contaminated aquifers is pump and treat; however, the presence of NAPLs decreases the ability of pump and treat systems to successfully reduce contaminant concentrations to acceptable regulatory levels (MacDonald and Kavanaugh, 1994). Since pump and treat systems rely on dissolution of the contaminant into the aqueous phase, the remediation time, and consequently remediation cost, is often inversely proportional to the aqueous solubility of the contaminant.

Surfactants have been shown to increase the aqueous solubility of organic contaminants and improve recovery from soil columns. Surfactant enhanced aquifer remediation (SEAR) may dramatically decrease remediation time and possibly reduce remediation costs. Initial field-scale studies of SEAR (GWRTAC, 1996; AATDF, 1997) have increased interest in this emerging technology; however, to date, there have been only a few economic analyses of SEAR implementations. Cost analyses utilizing a hypothetical site (AATDF, 1997) provide some insight into the economic constraints of SEAR, but these analyses lack a direct comparison of SEAR to pump and treat for a specific site. While it is possible to compare these costs to those for pump and treat as determined at other sites, this approach is not appropriate because each site has unique conditions and parameters.

The primary objective of this work was to evaluate and select surfactant formulations for implementation in the PCE source zone remediation at the Bachman Road site. Batch and 2-D aquifer cell experiments were conducted to assess PCE recovery and applicability of the two surfactant formulations. An additional objective was to develop and directly compare cost estimates for SEAR, based on micellar solubilization, and pump and treat for a PCE-contaminated field site.

Site Description. The Michigan Department of Environmental Quality (MDEQ) selected a PCE-contaminated aquifer located in Oscoda, Michigan for the evaluation of SEAR. Oscoda is located in northeastern Michigan on Lake Huron. The presence of dissolved phase organics was first detected in the water collected from residential wells along Bachman Road. Analysis showed the groundwater in the surrounding area to be contaminated with PCE, TCE (trichloroethylene) and BTEX (benzene, toluene, ethyl benzene and xylene). The source of the PCE contamination was determined to be a building previously occupied by N.S.I. Dry Cleaners. Dry cleaning operations have traditionally been a large user of PCE and evidence at the site, found during the Phase I site investigation, indicated the PCE source zone was located under the building. Phase I sampling data suggest that as much as 50 gallons (189 L) of PCE exist as an entrapped DNAPL. The PCE source zone, is located approximately 250 yards (229 m) west of Lake Huron with the groundwater flow in the direction of the lake. The depth to groundwater at the Bachman road site varies seasonally between 8-10 feet (2.4-3.0 m) and the depth to a clay-confining layer is approximately 25 feet (7.6 m). Thus, the aquifer thickness is 15-17 feet (4.6-5.2 m); however the site contains fine lenses, low permeability media located just above and approximately 15 feet (4.6 m) above the lower confining unit.

MATERIALS AND METHODS

Materials. Tween 80 (polyoxyethylene (20) sorbitan monooleate), a nonionic surfactant, was obtained from ICI Surfactants and was used as received. Aerosol MA-80I (sodium dihexyl sulfosuccinate), an anionic surfactant, was obtained from Cytec Industries and was used as received. Aerosol MA-80I contains 80% active ingredient and 20% inactive ingredients including water and isopropanol. HPLC-grade tetrachloroethylene (PCE), was obtained from Aldrich Chemical. PCE has an aqueous solubility of 150 mg/L and a density of 1.62 g/mL at 20°C. An organic soluble dye, Oil-red-O, was used at a concentration of 4×10^{-4} M to color the PCE for visualization purposes. Oil-red-O, HPLC-grade 2-propanol (IPA), calcium chloride and sodium chloride were obtained from Fisher Scientific.

A 4% (weight) solution of Tween 80 was prepared in water purified using a Nanopure Analytical Deionization (Barnstead/Thermolyne Corporation) system. The 4% Tween 80 solution also contained 500 mg/L $CaCl_2$ as a background electrolyte. Since Tween 80 is a nonionic surfactant, the addition of salt did not affect the phase behavior or the solubilization capacity of the surfactant solution.

Similarly, a solution of 8% (weight) Aerosol MA and 8% (weight) IPA was prepared with 15,000 mg/L NaCl and 1,000 mg/L CaCl$_2$.

Aquifer material, obtained from the Bachman Road site in Oscoda, Michigan, contained approximately 0.15% total carbon and 0.02% organic carbon, indicating a large fraction of inorganic carbon. The aquifer material was air dried and sieved to pass a #20 sieve. Ottawa F-70 (Ottawa 40-270 mesh) was obtained from U.S. Silica Co. and used as received. Ottawa F-70 represents approximately an order of magnitude permeability reduction from the Bachman aquifer material, and was used to create lenses of low permeability within the 2-D aquifer cell matrix.

Batch Methods. Completely mixed batch reactor experiments were conducted with each surfactant solution and PCE to determine the interfacial tension (IFT) and equilibrium PCE solubility. PCE was added to aqueous surfactant solutions, in 35 mL borosilicate glass centrifuge tubes in excess of the solubility capacity of the surfactants. The samples were mixed for 72 hours on a Lab Quake lab shaker and then the two phases were separated for analysis. The equilibrium concentration of PCE in each surfactant solution was determined by gas chromatography, and the IFT between PCE and each surfactant solution was measured using the drop volume method. The density of each surfactant solution as measured using 25 mL borosilicate glass pycnometers and dynamic viscosity was measured using a Haake RS75 RheoStress rheometer equipped with a double-gap cylinder sensor.

2-D Aquifer Cell Methods. A 2-D aquifer cell (60.75 cm x 40.32 cm x 1.38 cm) was constructed from parallel glass plates separated by square aluminum tubing. The end chambers of the aquifer cell (1.27 cm square aluminum tubing) were fully screened over the entire height of the cell. The cell was packed under saturated conditions with Bachman aquifer material. Low permeability lenses, consisting of F-70 Ottawa sand, were used to mimic regions of low permeability observed in field cores. One set of these lenses was packed up to 3 cm above the bottom of the cell, while a second layer of F-70 lenses was incorporated 19 cm above the bottom and 19 cm from either side of the cell. The final height of the Bachman material was 35 cm above the bottom of the cell. The packed cell was allowed to settle for at least 12 hours before initiating water flow. After this settling period, the permeability of the cell was determined by measuring the flow rate under a constant head condition.

Dyed PCE was injected into the water-saturated cell using a Harvard Apparatus syringe pump at a constant flow rate (0.5 mL/min). After the injection was stopped, the PCE was allowed to redistribute for 12 hours. Surfactant solution was then flushed through the box at a nominal flow rate of 3 mL/min and effluent samples were collected using an Isco fraction collector for gas chromatographic analysis.

Cost Analysis. The objective of the pilot-scale economic analysis was to compare the cost of the two SEAR alternatives to the cost of conventional pump and treat remediation. Two effluent water treatment systems, air stripping and activated carbon, were considered for the pump and treat system. Air stripper towers used in this study were designed using a 20% safety factor to account for the expected reduced mass transfer in the presence of the surfactant. The second water treatment system considered for the pump and treat system was liquid adsorption by granular activated carbon (GAC). Both pump and treat alternatives were compared to the two SEAR formulations for remediating the site to the maximum contaminant level (MCL) of PCE, 5 µg/L. Estimates for both SEAR alternatives included 1 week for set up of the system and to establish the hydraulic gradient, in addition to the time required to the flush of three pore volumes of water through the swept zone following surfactant injection.

Labor, equipment and material costs were obtained from Means (1999). The costs were separated into capital costs and operation and maintenance (O&M) costs. The salvage value of reusable equipment was obtained using the Modified Accelerated Cost Recovery System (MACRS) with an assumed 10-year class life for all depreciable equipment. Surfactant costs were obtained from three companies. ICI surfactants, manufacturer of Tween 80, supplied the cost of the Tween 80 as $1.35/lb. Cytec Industries, the manufacturer of Aerosol MA-80I, quoted a cost of $1.28/lb for the Aerosol MA-80I. Finally, Southchem, a chemical distributor in Greensboro, NC priced isopropanol at $0.27/lb. Although the surfactant solution costs were listed as capital cost, they are assumed to have no salvage value because the solution will be discharged to a sewer.

The total present value, assuming an interest rate of 5%, of each remediation alternative was calculated by adding the capital costs, surfactant solution costs, and O&M costs, and subtracting the salvage value. To this value, contingencies and fees were added, and a location factor of 0.95 was applied to arrive at the total present value.

RESULTS AND DISCUSSION

Experimental. Batch experimental results for density, interfacial tension and equilibrium solubility are summarized in Table 1. The weight solubilization ratios (WSR) of the Tween 80 and MA/IPA solutions were calculated to be 0.65 and 0.98 respectively. If the Tween 80 solution was increased in concentration to 8% active ingredient the equilibrium solubility of PCE would be approximately 52,100 mg/L compared to the value of 76,360 obtained for the 8% MA/IPA solution.

Two 2-D aquifer cell studies were performed to test the ability of each surfactant formulation to recover PCE from Bachman aquifer material. The pore volume, permeability and overall PCE saturation determined for each experiment are given in Table 2. For the Tween 80 experiment, a total of 7300 mL (~8 pore volumes) of the 4% Tween 80 + 500 mg/L $CaCl_2$ solution was delivered to the aquifer cell in four floods, with periods of flow interruption, approximately 12 hours in duration, introduced between each of the first three floods. The overall average flow rate for the Tween 80 experiment, excluding the

TABLE 1. Surfactant Solution Properties at 20°C

Solution	Density (g/mL)	IFT with PCE (dynes/cm)	Solubility of PCE (mg/L)	Viscosity (cP)
4% Tween 80 + CaCl$_2$	1.002±0.001	4.90±0.11 [1]	26,060±99	1.03±0.01
8% MA/IPA + NaCl + CaCl$_2$	1.009±0.001	0.160±0.004	76,360±2380	2.48±0.02
1. Taylor, 1999				

periods of flow interruption, was 3.19±0.01 mL/min which corresponded to a Darcy velocity of 7.22±0.70x10^{-2} cm/min. The experimental Darcy velocity was similar to the Darcy velocity proposed for the pilot-scale test, estimated to be approximately 8.56x10^{-2} cm/min.

TABLE 2. 2-D Aquifer Cell Experimental Parameters

Surfactant Formulation	Pore Volume (mL)	Overall Intrinsic Permeability (cm2)	Volume PCE (mL)	Overall PCE Saturation (%)
4% Tween 80	856	4.90E-07	41	4.8
8% MA/IPA	848	3.34E-07	27	3.2

Overall recovery of PCE for the 4% Tween 80 experiment was 53% after flushing 8 pore volumes of solution. The PCE mass remaining at the conclusion of the experiment existed in pools along the bottom of the cell. Because the pooled PCE had a low specific surface area, mass transfer of PCE was limited. The observed increase in effluent concentration after periods of flow interruption indicated the solubilization was also rate limited. Similar results have been observed in other studies (Pennell, *et al.*, 1993), indicating that rate limited micellar solubilization must be accounted for when developing flushing schemes and cost estimates for SEAR.

For the MA/IPA experiment, a total of 2450 mL (~3 pore volumes) of the 8% Aerosol MA/IPA + 15,000 mg/L NaCl + 1000 mg/L CaCl$_2$ solution was delivered to the aquifer cell in two floods, with a period of flow interruption between the two floods. The overall average flow rate, excluding the period of flow interruption, was 2.95±0.31 mL/min. This flow rate corresponded to a Darcy velocity of 6.68±0.70x10^{-2} cm/min which was again similar to the estimated pilot-scale Darcy velocity (8.56x10^{-2} cm/min). The 78% PCE recovery achieved with the MA/IPA solution was greater than the recovery using the Tween 80 solution (53%). Furthermore, the effluent concentrations of PCE after the flow interruption, suggest that the solubilization kinetics of the MA/IPA solution for PCE were faster than those observed for the Tween 80 solution.

Because surfactants reduce the IFT, which could lead to uncontrolled DNAPL mobilization, it is useful to calculate the total trapping number (Pennell, *et al.*, 1996) to predict if mobilization of free product will occur. Total trapping numbers were calculated for both the Tween 80 solution (6.1×10^{-5}) and the MA/IPA solution (1.3×10^{-3}). Since the total trapping number for the MA/IPA solution was greater than the critical value (1×10^{-5}), mobilization was expected and confirmed by visual observation. Although mobilization did occur in the MA/IPA experiment, no PCE was collected in the effluent as free product because a large fraction of PCE was solubilized prior to reaching the lower boundary of the aquifer cell. This observation is consistent with the large aqueous solubility of PCE ($76,360 \pm 2380$ mg/L) in the MA/IPA solution. While much of the mobilized PCE free product appeared to be solubilized, some PCE entered the lower fine lenses and confining layer.

Pilot-Scale Economics.   The pilot-scale flushing scheme consisted of six injection wells and one extraction well. The building under which the source zone is located provided an obvious physical limitation to the well locations. Three water injection wells were used to produce the necessary hydraulic gradient while maintaining hydraulic control of the surfactant. The time of travel, through the source zone, varied spatially, up to a maximum travel time (10 days). While the time to flush one pore volume of surfactant was assumed to be 2 days, the subsequent water flood must account for the travel time variation.   As a conservative estimate, the time to flush one pore volume of water through the source zone was taken to be 10 days, which allowed all surfactant to be recovered, regardless of position inside the pore volume.

The extraction well will be operated at a flow rate of 5 GPM (18.9 LPM) and consequently the effluent treatment system will be operated at the same flow rate.   The major equipment for the pilot-scale SEAR system consisted of 3 influent holding tanks, a mixing tank, 7 wells, an equalization tank, an air stripper and off gas carbon adsorption.   For the pump and treat system major equipment was based on the same process flow without the influent tanks and wells (pump and treat air stripping and carbon adsorption) and without the air stripper (pump and treat carbon adsorption).

The present value of each remediation alternative in terms of total cost, cost per volume soil treated, and total time are shown in Table 3. While the total remediation times for the surfactant solutions were substantially less than those for the pump and treat systems, the 8% MA/IPA solution operated for only 3 days less than the 4% Tween 80 solution. This does not appear to agree with the solubilities and recoveries observed in the experimental portion of this work, until one considers the overall saturation of the pilot-scale (0.4%) relative to that of the 2-D aquifer cell experiments (5%, Tween 80; 3%, MA/IPA). Because the exact distribution of PCE at the site was not determined, a uniform distribution was assumed. The assumption of uniform distribution led to the low overall saturation (0.4%) at the pilot-scale. Consequently, the low saturation required a relatively small volume of surfactant solution (2.9 pore volumes, 4% Tween 80; 1.3 pore volumes 8% MA/IPA), compared to the 2-D aquifer cell experiments (8 pore volumes, 4% Tween 80; 3 pore volumes 8% MA/IPA). These volumes correspond to flushing times of 6 days and 3 days for the Tween 80 and MA/IPA solutions respectively. The economic analyses were developed such that the surfactant flood was followed by 3 pore volumes of water, to displace all resident surfactant solution.

Based on the above analysis the more favorable pump and treat alternative utilized GAC for above-ground water treatment, with an estimated total cost of $307,000. Table 3 also indicates that remediation times were reduced substantially with the use of surfactants. The two surfactant systems evaluated, 4% Tween 80 + 500 mg/L CaCl$_2$ and 8% Aerosol MA/ IPA + 15,000 mg/L NaCl + 1000 mg/L CaCl$_2$, yielded similar costs with a total cost of $212,000 and $238,000 respectively. Often SEAR systems are assumed to be more expensive than pump and treat because of the large capital outlay at the beginning of the project; however, Table 3 shows that both the costs and remediation times of the surfactant systems were significantly less than the costs of the pump and treat systems.

Table 3. Pilot-Scale Costs

Alternative	Total Cost	$/m3 treated	Time
4% Tween 80	$ 212,000	$ 1,430	44 days
8% MA/IPA	$ 238,000	$ 1,600	41 days
Pump and Treat Air Stripping	$ 356,000	$ 2,400	1125 days
Pump and Treat GAC	$ 307,000	$ 2,070	1125 days

Surfactant recycle systems are often integral components in SEAR applications because of the high cost of surfactants. In the Bachman Road pilot-scale design, the use of a recycle system was not feasible because the volume of surfactant that would have been re-injected was insufficient to justify the cost of a recycle system.

CONCLUSIONS

Although no free product was collected in the effluent of the MA/IPA experiment, mobilization of the PCE was observed. The use of the MA/IPA solution in the field would pose substantial risk because there is no reliable way to insure that mobilized PCE will be entirely solubilized before entering lower permeability layers or lenses. This problem rendered the MA/IPA solution less attractive than the Tween 80 solution for implementation at the Bachman Road site. While the equilibrium solubility of the Tween 80 was less than that of the MA/IPA solution, the Tween 80 solution did not mobilize PCE. Thus, the 4% Tween 80 + 500 mg/L CaCl$_2$ formulation was selected based on the experimental results and similarity in cost.

The use of the Tween 80 should meet with regulatory approval easily, since it is food grade and biodegradable. Additionally, it was slightly less expensive than the Aerosol MA alternative and provides a large reduction in cost and time compared to pump and treat. The Tween 80 solution reduced the costs $95,000 as compared to the pump and treat GAC for the pilot-scale; furthermore, the reduction in time compared to the pump and treat systems was 1,080 days at the pilot-scale. These reductions are quite large and demonstrate that surfactants can be implemented economically for DNAPL source zone remediation.

ACKNOWLEDGEMENTS

We would like to thank ICI Surfactants and Cytec Industries for supplying the surfactants used in this work. Funding for this research was provided by the Great Lakes and Mid-Atlantic Center for Hazardous Substance Research under a grant provided by the State of Michigan Department of Environmental Quality (MDEQ). The content of this publication does not necessarily represent the views of the MDEQ and has not been subject to agency review.

REFERENCES

Advanced Applied Technology Demonstration Facility Program. 1997. *AATDF Technology Practices Manual for Surfactants and Cosolvents.*

Jaffert, C. T., Technology Evaluation Report. 1996. *Ground-Water Remediation Technologies Analysis Center (GWRTAC).*

MacDonald, J. A. and M. C. Kavanaugh. 1994. "Restoring Contaminated Groundwater: An Achievable Goal?" *Environmental Science and Technology.* 28(4): 362A-368A.

Pennell, K. D., L. M. Abriola, and W. J. Weber. 1993. "Surfactant-Enhanced Solubilization of Residual Dodecane in Soil Columns. 1. Experimental Investigation." *Environmental Science and Technology.* 27(12): 2332-2340.

Pennell, K. D., G. A. Pope and L. M. Abriola. 1996. "Influence of Viscous and Buoyancy Forces on the Mobilization of residual Tetrachloroethylene during Surfactant Flushing." *Environmental Science and Technology.* 30(4): 1328-1335.

R. S. Means Company. *Environmental Remediation Cost Data-Unit Price.* 1999.

Taylor, T. P., K. D. Pennell, L. M. Abriola and J. H. Dane. 1999. "Surfactant Enhanced Recovery of Tetrachloroethylene from a Porous Medium Containing Low Permeability Lenses. 1. Experimental." *Journal of Contaminant Hydrology.* (Submitted).

SURFACTANT-ENHANCED SUBSURFACE REMEDIATION OF DNAPLS AT THE FORMER NAVAL AIR STATION ALAMEDA, CALIFORNIA

Mark H. Hasegawa (Surbec-ART Environmental, LLC)
Bor-Jier Shiau (Surbec-ART Environmental, LLC)
David A. Sabatini (University of Oklahoma, Norman, Oklahoma)
Robert C. Knox (University of Oklahoma, Norman, Oklahoma)
Jeffrey H. Harwell (University of Oklahoma, Norman, Oklahoma)
Rafael Lago (Tetra Tech EM Inc., San Francisco, California)
Laura Yeh (Naval Facilities Engineering Service Center, Port Hueneme, California)

ABSTRACT: A surfactant enhanced subsurface remediation (SESR) field demonstration was conducted at Alameda Point (formerly Naval Air Station Alameda) to demonstrate enhanced dense non-aqueous phase liquid (DNAPL) removal at Site 5. The overall project goal was to remove 95% of the trapped DNAPL from fill sand within a 20' x 20' x 5' volume. The DNAPL at the site consists primarily of 1,1,1 Trichloroethane (TCA), Trichloroethylene (TCE) 1,1 Dichloroethylene (DCE), and 1,1 Dichloroethane (DCA). The surfactant solution selected for field application consisted of hexadecyl diphenyloxide disulfonate, (C16-DPDS or Dowfax 8390), sodium dihexyl sulfosuccinate (AMA-80), calcium chloride, and sodium chloride. Six pore volumes of surfactant solution were flushed through the test area over a 15-day period using four injection wells and four recovery wells and was followed by 9 days of water flooding. A macro porous polymer (MPP; Akzo Nobel) liquid-liquid extraction system was utilized to remove chlorinated solvents from the recovered surfactant solution. Surfactants were then concentrated using micellar enhanced ultrafiltration (MEUF) and re-injected into the test cell. Post-test soil coring results (8 soil borings and 18 samples) showed the chlorinated solvent concentrations to be non-detectable within the fill sand after surfactant flushing. Pre- and post-partitioning interwell tracer tests (PITT) were also conducted to evaluate NAPL removal. These results showed over 97% of DNAPL mass removal from the test cell area. Over 320 kg of Trichloroethane and Trichloroethene were removed from the cell during the 24-day period.

INTRODUCTION

A surfactant enhanced subsurface remediation (SESR) field demonstration was conducted by Surbec-Art Environmental L.L.C. (Norman, OK) at IR Site 5, Alameda Point, California (formerly Naval Air Station Alameda) in 1999. The U.S. Navy (EFA West) selected this surfactant flushing technique as one of the potential site remediation options in 1998. The selected test location is on the eastside of Building 5 at Site 5, where several chlorinated hydrocarbons (e.g., 1,1,1 trichloroethane and trichloroethylene) were found in the soil and groundwater samples. The purpose of this demonstration was to demonstrate enhanced removal of dense non-aqueous phase liquid (DNAPL) by flushing surfactant solution through the test site. This pilot-scale field demonstration was performed by Surbec-Art with support from Levine Fricke Recon (LFR) and The University of Oklahoma as subcontractors.

Technology performance was evaluated using three primary methodologies:
1. Pre- and post-partitioning interwell tracer tests (PITTs),
2. Pre- and post-soil coring, and
3. Mass recovery of DNAPL at the extraction wells (NOTE: this will cover any free-phase NAPL recovered, as well as dissolved phase contaminant, but is OK even if only dissolved phase was recovered).

The PITT is a technology that involved sweeping a suite of tracers through the subsurface to measure NAPL volume before and after the surfactant flood. The tracers were selected in order to provide a range of partitioning coefficients consistent with the NAPL being characterized, so that both large and small quantities of DNAPL can be detected within a reasonable timeframe. The partitioning of tracers into NAPL causes retardation in their arrival at the recovery wells compared to conservative or non-partitioning tracers which are also injected. The curves representing the breakthrough of the tracers in the recovery wells are used to quantify NAPL volume within the flushed zone. In this research, the PITT results were corroborated using pre- and post-soil coring analysis and mass removal of DNAPL in the recovered groundwater.

Background. SESR is a unique technology for expediting subsurface remediation of NAPLs. The surfactant system, usually an anionic or nonionic surfactant, is designed to remove organic contaminants (including chlorinated solvents) from contaminated soil. Surfactant/cosolvent systems can increase the solubility of hydrophobic organic compounds by several orders of magnitude and/or can significantly increase the mobility of NAPLs. The result can be a significantly reduced remediation time, increased removal efficiency (up to 3 or 4 orders of magnitude), and reduced cost of NAPL removal (Sabatini et al., 1998).

Surfactant injection solutions can be designed to be effective under most subsurface conditions. Multiple contaminants may pose a challenge, and naturally occurring divalent cations and salts can affect the performance of certain surfactants; however, it is possible to design an effective surfactant system for removal of the target contaminants under any of these conditions.

A key aspect of this project was to demonstrate contaminant separation from the surfactant solution for surfactant reuse. In order to re-inject the surfactant, contaminant concentrations must be reduced to acceptable levels in the surfactant solution and then the surfactant must be re-concentrated for reinjection. An example of the overall treatment process is illustrated in Figure 1. (Lipe et al., 1996; Hasegawa et al., 1997).

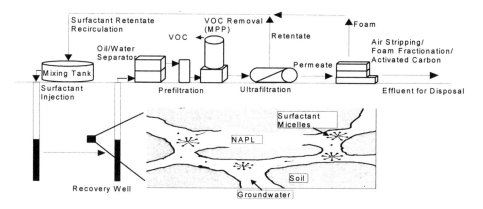

Figure 1: Process Flow Schematic

Objectives. The primary objective of the SESR project was to demonstrate the increased effectiveness of SESR for removing chlorinated solvent mass from the subsurface relative to conventional pumping techniques. The Navy established a DNAPL mass removal goal of 95% (from the soil) for this project. Three secondary objectives were to: 1) determine the physical properties of the porous media within the study area, including the hydraulic and sorption properties of the aquifer material, 2) determine the optimal surfactant mixture for DNAPL removal, and 3) determine the efficiency of surfactant recovery from the study area.

SURFACTANT SELECTION AND SYSTEM DESIGN
Surfactant Selection. Site soil and groundwater samples were obtained and used to screen for the optimal surfactant system. In the laboratory screening experiments, eight anionic surfactants or mixtures were investigated for their potential use in remediating TCA, TCE, DCE, and DCA or their mixtures with the in situ surfactant flushing technology. These surfactant systems were selected based on their previous use, cost, toxicity, and availability. The laboratory screening activities consisted of numerous tests such as contaminant solubilization, surfactant sorption and precipitation, surfactant–NAPL phase properties and contaminant extraction-column studies. Based on the screening tests, we concluded that the following surfactant system was the best candidate for this test: Dowfax 8390 (5 wt%), AMA-80 (2 wt%), NaCl (3 wt%), CaCl$_2$ (1 wt%).

Groundwater Treatment System Design: As part of the design process, site characterization data was input into in a three-dimensional numerical groundwater model (Visual Modflow) to design the injection/recovery system and predict production rates for process design. A line drive injection/recovery system with a total production of 8 gpm (2 gpm per recovery well; refer to Figure 2) was selected as optimal for application at this site.

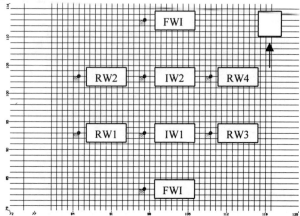

Figure 2: Well Location Map (RW: recovery well; IW: surfactant injection well; FWI: freshwater injection well for hydraulic control)

The (effluent treatment or surfactant recovery) process equipment consisted of a trailer-mounted oil/water separator, pre-filtration, micellar-enhanced ultrafiltration (MEUF), and tray stripper, as well as a separate skid-mounted macro porous polymer liquid-liquid extraction system (MPP, Akzo Nobel). Both were designed using the breakthrough curves predicted by the numerical modeling simulations (see Figure 1).

The MEUF system consisted of three spiral-wound cross flow filters: 2-4" x 4' and 1-8"x4'. A 5 hp centrifugal pump, capable of attaining 60 gpm at a pressure of 60 pounds per square inch (psi), was used to feed solution into the MEUF. The filters were 10,000 molecular weight (MW) cutoff ultrafiltration membranes purchased from Koch Industries.

IMPLEMENTATION
Design of the PITT. Interwell partitioning tracer tests (PITTs) were used to evaluate NAPL distribution in the test area, quantify pre-test NAPL concentrations in the injection zone, and target delivery of surfactant during remediation. The fundamental theory for designing the PITT for the characterization of NAPL can be found at great detailed in the literature (Jin et al., 1995). Selection of the pre- and post-partitioning tracers was finalized during lab-scale testing and based on their partitioning coefficients between the DNAPL and source water and the residual DNAPL concentration in the test cell.

Of the five tracers used during the pre-PITT, bromide and methanol were the conservative tracers. Hexanol, pentanol and 2,4-dimethyl-3-pentanol (DMP) were the partitioning tracers. The initial tracer concentrations used ranged from 694 mg/L to 1767.3 mg/l. The tracers used in the post-PITT included methanol, hexanol, DMP and heptanol. The initial tracer concentrations ranged from 510 mg/L to 1708 mg/l. The total volume of tracer solution injected was 1,000 gallons in both pre- and post-PITT. The tracer samples were analyzed by the GC/FID method (see Surbec-Art, 1999 for detailed).

Surfactant Flooding. Injection of the surfactant solution into IW-1 and IW-2, and fresh water into FWI-1 and FWI-2 began on August 9, 1999. The collected surfactant samples were analyzed using a HPLC system with dual UV and conductivity detectors. The VOC samples collected were analyzed onsite using the GC/FID method using a pre-packed FS wool inlet liner (Resteck, PA) to remove the interferences of surfactant.

During the surfactant flood, recovered groundwater was piped through the oil/water separator, bag filters (prefiltration), MPP, and MEUF (refer to Figure 1). Permeate from the MEUF was sent to a 21,000-gallon holding tank while retentate was collected in the mixing tank. The retentate solution was stored in a 3,000-gallon tank where surfactant solution constituents were monitored and adjusted to appropriate reinjection levels before re-injection. A total of 84,302 gallons of surfactant solution was injected during the surfactant flood and 97,552 gallons of groundwater was recovered. During the SESR portion of the test, 1,540 gallons of Dowfax 8390, 1,045 gallons of AMA 80, 3,440 pounds of $CaCl_2$, and 5,120 pounds of NaCl were mixed in the 84,302 gallons of injected solution.

After the surfactant flood fresh water was injected to flush the surfactant solution out of the cell. During the fresh water flood, 37,141 gallons of water were injected and 41,083 gallons were recovered. Treated permeate was piped to the East Bay Municipal Sewer District for Disposal.

RESULTS AND DISCUSSION
Contaminant Breakthrough and Recovery. Key contaminants detected at the site in declining order are TCA, TCE, DCA and DCE. Among these, TCA and TCE are the parent products that were originally spilled and are the main constituents of the DNAPL collected (over 80%). As a result, the data presented illustrates the combined concentrations of TCA and TCE. The contaminant breakthrough curves and the mass recovered from the recovery wells are shown in Figures 3 and 4, respectively.

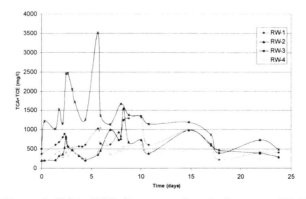

Figure 3: TCA+TCE Concentrations in Recovery Wells

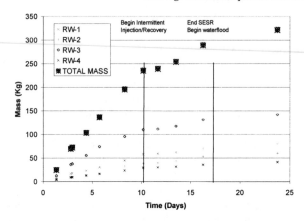

Figure 4: Mass TCA+TCE Recovery from Wells

During the surfactant flood most of the contaminant mass was recovered from RW-3 (southeast quadrant) followed by RW-1, RW-2 and RW-4 (Figure 3). Based on the flow rates and concentrations in the recovery wells, these results indicate that the total mass of TCA and TCE recovered was 320 kg (65 gallons; refer to Figure 4). It should be noted that total amount of NAPL recovered should be greater than 65 gallons; or, approximately 80 gallons, if DCE and DCA were included. The mass recovery trend is a steep linear relationship through day ten (four pore volumes), followed by a significant reduction in slope. This is consistent with the depletion of DNAPL as the project proceeds. The graph also illustrates the rate of mass recovered from each well. The mass recovery and depletion of DNAPL trends are similar despite the differences in absolute values between wells.

PITT Results. The mass of DNAPL present within the demonstration cell was estimated using two techniques, PITTs and soil coring. An example of tracer breakthrough curves is depicted in Figure 5 for RW-1 during the pre-PITT. The regression shown in the figure aids in defining the tail of each curve. Based upon the pre-PITT calculations, the estimated volume of DNAPL in the cell ranged from 100 to 169 gallons of DNAPL, as summarized in Table 1. This data was estimated by calculating the first moment of each breakthrough curve.

Results of post-PITT indicated that the NAPL volume left in the cell after surfactant flood ranged from 0 to 3 gallons. The swept pore volumes measured before and after surfactant flood were similar (8499 gals vs. 8677 gals). Based on the PITTs, it is estimated that more than 99% of the DNAPL present in the demonstration cell prior to implementation of the surfactant injection was removed during the surfactant injection and recovery.

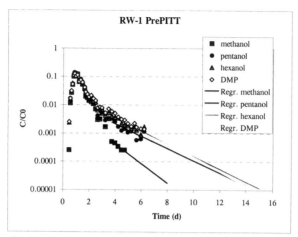

Figure 5: Pre-PITT Breakthrough for RW-1 (C/C0 vs. Time)

Table 1. Results from PITT

	RW-1	RW-2	RW-3	RW-4	Total
Pre-Test NAPL Volume (gal)					
Hexanol	30	53	29	57	**169**
DMP	16	32	21	31	**100**
Post-PITT NAPL Volume (gal)					
DMP	0	0	0	0	**0**
Heptanol	3	0	0	0	**3**

Soil Coring Results. The mass of DNAPL present in the cell was also estimated based on the soil coring results obtained during installation of the well network. Soil corings were taken before and after the surfactant flood. The cell was divided into quarter sections and 18 samples were taken from eight soil borings distributed evenly between the sections. The volume of DNAPL was estimated by averaging a mass for each section. The total mass was converted to a volume of 34 gallons (before) and less than one gallon (after) using an assumed density of the mixed DNAPL. It is estimated that more than 97% of the DNAPL present in the demonstration cell prior to implementation of the surfactant injection was removed during the surfactant injection and recovery.

Surfactant Recovery. A total of 1540 gallons of Dowfax and 1045 gallons of AMA were used to prepare the injection solutions. Since the actual mass of surfactant injected into the subsurface included surfactant recovered using the MEUF system, the total mass of surfactant injected was actually higher. The total mass recovery of Dowfax and AMA was 117% and 90%, respectively, as compared to the surfactant mass injected.

Volatile Organic Compounds (VOC) Removed by the MPP System. The removal of TCA and TCE by the MPP was less effective while surfactant concentrations were highest, during the surfactant injection. On the average, the MPP removed 80% (during peak surfactant concentrations) to 95% (lower surfactant concentrations) of the contaminant from the waste stream. Overall, the MPP system provided an effective, trouble free performance and is a viable technology for contaminant separation from solution.

CONCLUSIONS

Pre-test results suggest that there are between 34 gallons to 169 gallons of DNAPL present in the subsurface. This variation reflects a number of difficulties in quantifying DNAPL volumes within the subsurface. First, the width between soil borings is usually too great to provide an accurate picture of site conditions. Second, DNAPL pooling or preferential flow paths can affect the sensitivity of PITT results. The pre-PITT results showed fairly substantial variations and the precision to assign to this analysis is still not clearly defined (Knox et.al., 2000). The post PITT and post soil coring results consistently showed little to no DNAPL remaining within the test cell.

The results of this project indicate that SESR will be effective for DNAPL mass removal at the subject site. Over 320 kg of TCA and TCE were recovered using four recovery wells. Soil and groundwater concentrations within the demonstration cell decreased significantly. Using the most conservative estimates both the PITT and soil coring results showed that DNAPL removal efficiencies exceeded the 95% target removal. The surfactant system (Dowfax, AMA, NaCl, CaCl) significantly enhanced the DNAPL solubility and extraction efficiency. The MPP efficiently decontaminated the surfactant solution for reinjection during the demonstration. This demonstration project thus illustrated the effectiveness and robustness for remediating subsurface DNAPL contamination at the subject site.

REFERENCES

Jin, M., M. Delshad, V. Dwarakanth, D. C. McKinney, G. A. Pope, K. Sepehrnoori, C. Tilburg, and R. E. Jackson. 1995. "Partitioning tracer test for detection, estimation and remediation performance assessment of subsurface nonaqueous phase liquids." Water Resources Research. 31(5): 1201-1211.

Knox, R., Goodspeed, M., Sabatini, D., 2000. "Partitioning Interwell Tracer Tests as an Assessment Tool For DNAPL Quantification", In preparation,

Lipe M. D. Sabatini, M. Hasegawa, and J. Harwell. 1996. *Ground Water Monitering and Remediation.* 16(1): 85-92.

Hasegawa M., D. Sabatini, and J. Harwell. 1997. *Journal of Environmental Engineering Division, ASCE.* 123(7): 691-697.

Sabatini, D. A., J. H. Harwell, and R. C. Knox. 1998. "Surfactant selection criteria for enhanced subsurface remediation: Laboratory and field observations." Progr. Colloid Polym Sci. 111: 168-173.

2000 AUTHOR INDEX

This index contains names, affiliations, and book/page citations for all authors who contributed to the seven books published in connection with the Second International Conference on Remediation of Chlorinated and Recalcitrant Compounds, held in Monterey, California, in May 2000. Ordering information is provided on the back cover of this book.

The citations reference the seven books as follows:

2(1): Wickramanayake, G.B., A.R. Gavaskar, M.E. Kelley, and K.W. Nehring (Eds.), *Risk, Regulatory, and Monitoring Considerations: Remediation of Chlorinated and Recalcitrant Compounds.* Battelle Press, Columbus, OH, 2000. 438 pp.

2(2): Wickramanayake, G.B., A.R. Gavaskar, and N. Gupta (Eds.), *Treating Dense Nonaqueous-Phase Liquids (DNAPLs): Remediation of Chlorinated and Recalcitrant Compounds.* Battelle Press, Columbus, OH, 2000. 256 pp.

2(3): Wickramanayake, G.B., A.R. Gavaskar, and M.E. Kelley (Eds.), *Natural Attenuation Considerations and Case Studies: Remediation of Chlorinated and Recalcitrant Compounds.* Battelle Press, Columbus, OH, 2000. 254 pp.

2(4): Wickramanayake, G.B., A.R. Gavaskar, B.C.Alleman, and V.S. Magar (Eds.) *Bioremediation and Phytoremediation of Chlorinated and Recalcitrant Compounds.* Battelle Press, Columbus, OH, 2000. 538 pp.

2(5): Wickramanayake, G.B. and A.R. Gavaskar (Eds.), *Physical and Thermal Technologies: Remediation of Chlorinated and Recalcitrant Compounds.* Battelle Press, Columbus, OH, 2000. 344 pp.

2(6): Wickramanayake, G.D., A.R. Gavaskar, and A.S.C. Chen (Eds.), *Chemical Oxidation and Reactive Barriers: Remediation of Chlorinated and Recalcitrant Compounds.* Battelle Press, Columbus, OH, 2000. 470 pp.

2(7): Wickramanayake, G.B., A.R. Gavaskar, J.T. Gibbs, and J.L. Means (Eds.), *Case Studies in the Remediation of Chlorinated and Recalcitrant Compounds.* Battelle Press, Columbus, OH, 2000. 430 pp.

2000 KEYWORD INDEX

This index contains keyword terms assigned to the articles in the seven books published in connection with the Second International Conference on Remediation of Chlorinated and Recalcitrant Compounds, held in Monterey, California, in May 2000. Ordering information is provided on the back cover of this book.

In assigning the terms that appear in this index, no attempt was made to reference all subjects addressed. Instead, terms were assigned to each article to reflect the primary topics covered by that article. Authors' suggestions were taken into consideration and expanded or revised as necessary to produce a cohesive topic listing. The citations reference the seven books as follows:

2(1): Wickramanayake, G.B., A.R. Gavaskar, M.E. Kelley, and K.W. Nehring (Eds.), *Risk, Regulatory, and Monitoring Considerations: Remediation of Chlorinated and Recalcitrant Compounds.* Battelle Press, Columbus, OH, 2000. 438 pp.

2(2): Wickramanayake, G.B., A.R. Gavaskar, and N. Gupta (Eds.), *Treating Dense Nonaqueous-Phase Liquids (DNAPLs): Remediation of Chlorinated and Recalcitrant Compounds.* Battelle Press, Columbus, OH, 2000. 256 pp.

2(3): Wickramanayake, G.B., A.R. Gavaskar, and M.E. Kelley (Eds.), *Natural Attenuation Considerations and Case Studies: Remediation of Chlorinated and Recalcitrant Compounds.* Battelle Press, Columbus, OH, 2000. 254 pp.

2(4): Wickramanayake, G.B., A.R. Gavaskar, B.C.Alleman, and V.S. Magar (Eds.) *Bioremediation and Phytoremediation of Chlorinated and Recalcitrant Compounds.* Battelle Press, Columbus, OH, 2000. 538 pp.

2(5): Wickramanayake, G.B. and A.R. Gavaskar (Eds.), *Physical and Thermal Technologies: Remediation of Chlorinated and Recalcitrant Compounds.* Battelle Press, Columbus, OH, 2000. 344 pp.

2(6): Wickramanayake, G.B., A.R. Gavaskar, and A.S.C. Chen (Eds.), *Chemical Oxidation and Reactive Barriers: Remediation of Chlorinated and Recalcitrant Compounds.* Battelle Press, Columbus, OH, 2000. 470 pp.

2(7): Wickramanayake, G.B., A.R. Gavaskar, J.T. Gibbs, and J.L. Means (Eds.), *Case Studies in the Remediation of Chlorinated and Recalcitrant Compounds.* Battelle Press, Columbus, OH, 2000. 430 pp.